'What's striking is Peter's consu[...] ability to find radical answ[...] [...] that have confounded Britain's brightest minds' *Daily Telegraph*

'Life-affirming . . . the most extraordinary thing about scientist Dr Peter Scott-Morgan, aside from his optimism, is his relentless capacity for joy' *Daily Mail*

'A soaring love story' *Financial Times*

'Breathtaking . . . we were mesmerized by the boundless creative intelligence, positivity and love that pours from these pages. Profoundly inspiring' *Attitude*

'The astonishing true story behind the primetime Channel 4 documentary about Peter Scott-Morgan, the first person to combine his very humanity with artificial intelligence and robotics to become a full cyborg. His discovery means that his terminal diagnosis is negotiable, something that will rewrite the future – and change the world' *GQ*

'Ultimately this is a book about the very essence of what makes us human, of great courage and inspiration in the face of adversity, of hope and most importantly love' Paul Welham, Chairman and Chief Executive of CereProc, creator of text-to-speech solutions

'An extraordinary manifesto for a future where humans, AI and robotics collaborate rather than compete. Full of humour, pathos, challenging ideas and hope, *Peter 2.0* explores the fascinating philosophical dilemmas that such a future will involve' Marcus du Sautoy, Professor for the Public Understanding of Science at the University of Oxford and author of *The Creativity Code*

'Peter's extraordinary autobiography reminds us of what it is to be human and how important it is to live a life full of love. It reminds us that adversity can foster tenacity, and faced with the most insurmountable challenges we can find the strength to keep on fighting' Iain Canning, Academy Award-winning film producer best known for *The King's Speech* and *Lion*

'This is the story of what happens when one man harnesses intelligence, scientific insight and dogged determination to change the course of his life. It is a book filled with humour and lashings of bravery. But above all it is a love story! It is about the sort of love we all dream of, a love that conquers all, and that just might be truly eternal' Michele Romaine, journalist and broadcaster

'This is not a tragic story, but one of rebellion, of possibility and of a life of adventure just beginning . . . a thrilling account of human transformation' Baroness Jane Campbell of Surbiton, DBE

'A remarkable account of what it means to be human and what technology can really achieve' *Sunday Telegraph*

'Fascinating and extremely moving' *Sun*

ABOUT THE AUTHOR

Dr Peter Scott-Morgan is a robotics scientist who is the world expert on Unwritten Rules – civilization's behavioural algorithms that influence much of our future.

He has been granted unparalleled confidential access across North America, Europe, Asia-Pacific and Latin America to decode the hidden system dynamics driving economies, institutions, governments and corporations, including many of the major players in banking, chemicals, energy, healthcare, IT, manufacturing, media, retail, telecoms and transport.

Peter passionately believes the highest purpose of science is to help us all thrive, whatever we are, whatever our background, whatever our circumstances, whatever our ambitions.

He lives with his husband, Francis, in Torquay, England.

Peter 2.0

PETER SCOTT-MORGAN

PENGUIN BOOKS

PENGUIN BOOKS

UK | USA | Canada | Ireland | Australia
India | New Zealand | South Africa

Penguin Books is part of the Penguin Random House group of companies
whose addresses can be found at global.penguinrandomhouse.com

First published by Michael Joseph 2021
Published in Penguin Books 2022
001

Copyright © Peter Scott-Morgan, 2021

The moral right of the author has been asserted

Typeset by Jouve (UK), Milton Keynes
Printed and bound in Great Britain by Clays Ltd, Elcograf S.p.A.

The authorized representative in the EEA is Penguin Random House Ireland,
Morrison Chambers, 32 Nassau Street, Dublin D02 YH68

A CIP catalogue record for this book is available from the British Library

ISBN: 978–0–241–44710–9

To Francis, who married a cyborg

FRANCISCUS♥

PETRUSQUE♥C

ONTRA♥MUN

DUM♥CONIUN

CTI♥VINCEMUS

Acknowledgements

I decided to cheat with this book. It's an ego thing. In my defence, although I've written lots of books before, I've never met somebody who's actually read one. At least not somebody who's family or a close friend. At least not for fun. Or all the way through. So this time there's loads of kith and kin in the book itself, and I'm certainly not going to make the rookie mistake of letting on to them here who's been included. What they do next is basically unwritten rules. In fact, now I think about it, consider this not so much cheating as a sociology experiment.

An added bonus is I can avoid this Acknowledgements section turning into the usual Academy Awards acceptance speech; instead, I'll jump straight to the emotional crescendo near the end. Let me introduce two people and their teams without whom you and I would never have met: the amazing Rosemary and the equally amazing Dan.

Rosemary Scoular is my agent, a leading light at the enormous United Agents.

Now the thing you need to understand about agents is that a disconcerting percentage resemble aggressive rottweilers. Female agents are often the scariest – they're rottweilers in heels and lipstick. In which case, all I can say is that if Rosemary does indeed conceal any canine DNA then it's most likely cloned from a loyal terrier. The sort that won't let go even whilst affectionately wagging her tail. In short, Rosemary is absolutely lovely. And staggeringly effective. As are all her team. And in my resolute optimism I dare to hope that, long after I am of no further professional use to her, we will nevertheless be loyal friends. And metaphorically wag our tails when we see each other.

Daniel Bunyard is my publisher, a leading light at the enormous Penguin Random House.

Now the thing you need to understand about publishers is that I've worked with them my entire adult life. But in all that time I have never – ever – worked with one as compassionate, as intelligent, as generous, as insightful, as supportive, as courageous, as creative, as stimulating, as open to radical ideas. In short, as gobsmackingly good at his job. As a case in point, there are three chapters (you'll know them when you read them) that felt to me like a huge gamble. I was passionate about them, but they broke the rules, they broke new ground. I fully expected them to break Dan's and my cosy relationship. Instead, he was immediately as passionate about them as I was. And I am forever grateful. For everything.

Subsequently working with Dan's team was an equally joyous experience. It turned out to be a true ensemble of creative professionalism and prodigious talent that I have never before experienced in the publishing world. In truth, I have never had so much fun writing. Anything.

Now, for my penultimate thanks, I will break my embargo on family. I have three wonderful nephews – Lee, David and Andrew – who in different ways each help me thrive. But only Andrew can truthfully claim to have written half this book – albeit when I could no longer type and resorted to dictation.

Yet my ultimate thanks go to someone you will get to know very well. The unsung hero of this book. The true star. You will not yet fully understand the significance of what I now write but suffice to say that, when tested to the limits of endurance, the *real* Avalon proved himself to be infinitely braver, more loyal, more heroic, than Teenage Peter could ever have imagined . . .

Peter 2.0, Torquay, England, 2020

Peter's First Rule of the Universe

SCIENCE
is the only route to
MAGIC

The End

'So, what interminable string of letters have you built up after *your* name these days?' Anthony asked. He was being kind; I'd just been asking about the OBE after his.

'PhD, DIC, BSc(Eng), CITP, CEng, MBCS, MRTS, ACGI,' I gabbled, unthinkingly reverting to Schoolboy Peter and trying to show off to my oldest friend. Adult Peter attempted to retrieve the situation: 'But OBE is infinitely superior.'

Anthony and I had carefully coordinated our diaries so that we'd pass through London at the same time. He had an unbelievably full international schedule as head of the Lyric Opera in Chicago. Francis and I had spent much of the time since our marriage almost a decade ago exploring the world and visiting all the exotic locations we'd long dreamed of visiting. Today, however, Anthony and I had managed to find a day we could spend together, catching up and reminiscing.

'So, as it turned out, the historic occasion I missed out on was not your civil partnership but your marriage?'

'Yes! Last year they changed the law again and made it possible to upgrade. But it didn't actually matter because the legislation was retrospective; our marriage certificate was dated 21 December 2005. We'd been married for almost nine years without knowing it!'

'And it was great that both your parents lived to see it.'

They'd died within three months of each other – both in their late nineties. Francis and I had nursed them through their last two years.

'Now, tell me about your honeymoon in India. Colin and I plan to go there soon, so I want to pick up pointers!' Colin was Anthony's husband – they'd been together for decades. 'Was it on that trip you both developed your wanderlust?'

'It was, yes. One day we suddenly said: "Why don't we just keep travelling?" We'd saved heavily, and I know you can never have too much money, but you *can* have enough. Whereas, we realized, the one thing Francis and I'll never have enough of is time together while we're still relatively young. So I' – I tweaked my fingers symbolizing inverted commas – '"retired".'

'In your forties –'

'Well, just!'

I had still taken on the odd project working as a consultant to various companies and organizations, something I hugely enjoyed.

'And you're obviously loving "retirement"!' He mimicked my air quotes.

'It's perfect! I've written a couple more sciencey books, but I'm basically making up for all the geography and history and art that I had to give up at school. Francis and I are both fit and healthy. We both love the adventure. We plan to keep exploring for at least another couple of decades. At last, everything is right with the world . . .'

After a lovely hot soak in a bath far north of the Arctic Circle, on a winter trip in search of the aurora borealis, I had just stood up in the still-warm water and dried myself down to my knees. I lifted my left leg, shook off the drips – much like a dog that has stepped in something – set my foot down on the bath mat, then lifted my right leg. And that was when I juddered across what I can only surmise was a major discontinuity in the space–time continuum, after which I found myself hurtling toward an utterly foreign future.

My right foot didn't wiggle properly. It wiggled a bit. But it was far more like how a really ancient giant tortoise on the Galapagos might shake a leg. At best, it was a slow waggle. I casually noted the fact with a scientist's unending curiosity in all things that don't quite fit and got out of the bath anyway.

After it had happened a few more times, I concluded I had a wonky foot. Probably a pulled muscle. No biggy. With a perfectly reasonable working hypothesis lodged away in my subconscious, my brain relaxed. For a full three months. Until one day, when climbing up to a beautifully preserved Ancient Greek temple on the island of Rhodes, I started to vibrate.

Nothing dramatic, I hasten to add. But a definite occasional tremor in my right leg if I happened to move or sit down in a particular way. Sometimes. Hardly noticeable. A fortnight later, I consulted a physiotherapist about my wonky foot. He prodded and stretched and took lots of notes. Yes, it could well be a deep muscle strain. Maybe a slight tear. Any other symptoms? I mentioned the tremor.

'I didn't notice any –'

'No, you need to set it off, look, like this . . .'

'Aah.'

It was one of those mildly disconcerting aahs that suggest you don't necessarily want to know what is going through the mind of the person that said it. Of course, I *did*.

'So, when someone presents like that, what is it typically indicative of?' Intuitively, I'd switched to a more professional mode of speech. Depersonalize a sensitive situation. Discuss as a peer. Talk in the abstract. You learn more, faster, that way.

'Well, classically it's a sign of something called clonus.'

I couldn't remember ever having heard the term. And I'd vaguely memorized thousands of medical terms. Rather arrogantly, I automatically assumed it must be relatively uncommon. Without thinking, I hazarded: 'Nerve damage?'

'Exactly! It's suggestive of an upper motor neurone lesion. I'll write a letter to your GP immediately so he can refer you to a neurologist who can get you an MRI.'

As I walked out of my physio's office, clutching his hastily prepared letter to my GP like a talisman, I ran through the more obvious types of 'lesion' that could be affecting my spinal cord or brain (which was what my physio's phrasing had meant). Physical injury was an obvious one. But I hadn't been knocked about much since my late teens, when I did karate. Why would any damage only show now?

Cancer, of course. A brain tumour could do it, but I'd no other symptoms. A localized tumour on the spine? I wondered how operable that would be. A minor stroke? Not good news if it were the harbinger of more to follow. Not cerebral palsy – typified by uncontrolled movements – because that always started in childhood. But MS, multiple sclerosis? Pretty common. Could appear later in life. Incurable. Must be a contender. Still, better than a brain tumour.

Ten days later, I was lying flat on a narrow platform that was slowly sliding me backwards into the doughnut of a high-powered MRI machine. I'd never actually had an MRI scan before, so I was enthralled by the equipment. Magnetic resonance imaging uses superconducting magnets cooled by liquid helium that are so immensely powerful they can get parts of your body to emit tiny signals that can be processed to build a 3-D image. MRIs are also exceptionally noisy.

Even through the heavy-duty ear-defenders I'd been issued, the word 'cacophony' didn't come close. Huge magnetic forces thump on and off several times a second, causing all the equipment to pound in unison. Given that you are by this stage inside said equipment, you get pounded too. The quiet bits sound like someone trying to break into a metal helmet over your skull using a pneumatic hammer; the loud bits sound (and feel) like a full-on artillery bombardment. This is medical imaging shock-and-awe style. I was booked in for a couple of hours of it.

I've always felt I was good at reading people but my neurologist was giving away nothing. I recognized this for what it was: the habitual façade of someone who often has to break bad news. He indicated a chair for me facing a large computer screen, went to a corner of the room to fetch a chair for Francis, waited for us both to sit, sat himself, then suddenly smiled.

'Before I take you through the scan results, I just want to reassure you we haven't found anything nasty. So, you can relax!'

I realized I'd been holding my breath. My own professional veneer clicked in as I felt the prickling numbness of the huge dose of adrenaline that Peter the Scientist registered must have been released into my bloodstream only moments earlier in anticipation of The Verdict.

'Oh, well, that's very encouraging.' Calm. Low-key. As if he'd just told me that his prize-winning petunias were looking good again this year. Then intense curiosity, my lifetime companion, felt entitled to nuzzle its way to the fore.

'So, what on Earth is going on?'

My neurologist took us through my brain.

'You have a lovely brain,' he said with a pride that suggested part-ownership. 'You see ... within the skull ... nothing there at all ...' I wasn't sure that this was a quote that would play well as an endorsement on the cover of any book that I wrote, but I knew it was meant kindly.

There, slowly, all the way down my spinal cord, a reassuringly uniform dark circle surrounded by the grey and white abstract art of the undulating outlines of my vertebrae as we descended. Nothing untoward other than some previously unrecognized mild scoliosis (a sideways curve in the spine) that finally explained my unanswered teenage question of why one side of my pelvis was ever so slightly higher than the other.

In conclusion, everything was clear. No brain tumour. No spinal tumour. No signs of MS. No motor neurone disease. Not even the hint of a pinched nerve. All completely clear. Having ruled out all the obvious suspects, my neurologist advised my GP that I appeared to have something exotic. I was accordingly sent on a treasure hunt of increasingly esoteric tests.

We started with a mundane chest X-ray, but rapidly escalated to a large-page à la carte menu of blood tests. Translating the innocuous labels on the form, I noted that three tests alone were for AIDS-related infections – which was perfectly reasonable because autoimmune diseases can sometimes mimic neurological problems.

When all the results came back negative, we upped our game. This time, when the hospital phlebotomist saw the list, she couldn't stop herself saying: 'Oh, I do hope you don't have any of these. They're all horrible!' However, when these too all came back negative, the requested bloods appeared to migrate from the esoteric to the surreal. Still nothing. Soon, having run some genetics tests (negative), my neurologist seemed at a loss. With what I interpreted as poorly concealed desperation, he came up with what he assured me was his final list. As a result, my new best friend the phlebotomist was now also at a loss.

'I've never even heard of most of these.' She checked on

her computer listing of available tests. Two weren't even options. After several phone calls, she eventually got through to some obscure laboratory somewhere and sent them scurrying through their manual filing cabinets to uncover the information she needed.

'Well, at least we should finally get to the bottom of what's wrong with you.'

I really hoped so. I'd been monitoring my condition carefully. As every week passed, whatever was wrong with me was slowly but inexorably spreading up my leg. And until we got a diagnosis there was absolutely nothing that any of us could do to stop it. A fortnight later, the results came back.

Sweet Sixteen

It's the birthright of every human being to be able to change the universe.

Although I'd already reached this conclusion by my sixteenth birthday and liked the idea that breaking rules might be an indispensable factor in achieving it, I was still fuzzy on the details. That said, if I'd never intuitively rebelled a month later, on a gloriously sunny Wednesday afternoon in May, I wouldn't be alive today.

So I ought to start by giving effusive thanks to my old headmaster – an ironic turn of events given that I never rated him, either as a teacher or as a man. He, meanwhile, viewed me as an abomination against God and humanity. Nevertheless, credit where it's due: without his intervention that day I might by now be a fully paid-up member of the Establishment. I'd also be dead.

The significant action of that pivotal afternoon got going around 12.28, near the end of an English period. I was standing behind my rather ancient desk, pouring out my thoughts to my fellow pupils as I read them my essay on 'The Future'.

My rather ancient master was smiling and nodding encouragingly. Happily, he'd already marked my work 'A+' with the glowing commendation: 'Fantastical imaginings; thank goodness your prognostications are whimsy!' This slightly overlooked the point that I'd written my essay as a work of non-fiction, not fiction. To be fair, my mother may have subsequently made the same mistake because, of all the essays I ever brought home, after she'd died I found she'd kept this one.

'. . . my brain will link with an electronic brain and together we will be far more intelligent than the sum of our parts. In conclusion, my five –'

The bell rang to signal lunchtime. No one moved – the bell was a guide for masters not pupils. With a flick of his wrist the master indicated I should continue.

'In conclusion, my five senses will be augmented by myriad electronic components until my whole identity will evolve, my humanity will evolve. Instead of driving a car, or even a huge ship, I will *be* that car; I will *be* that ship.' I looked up at the master to confirm I had finished.

'Wonderful imaginative caprice, Scott. You'll be promulgating science fantasy yet! Very good.' He looked at the expectant class. 'Dismissed!'

'You are *such* a crawler!' Simpson, getting up from behind the desk to my left was not always my greatest fan. 'You stole all of that from *Doctor Who.*'

As I stashed my things away in my desk, I felt the need to counter his slander: 'Actually, since 5.15 p.m. on Saturday 23 October 1963, there has never been a single episode of *Doctor Who* that even vaguely corresponded to the extrapolation you have just heard.'

'Honestly?' This was slim little Connor, bizarrely a friend to both Simpson and me, who had joined us en route to the door of the classroom. I liked Connor. A lot.

'You really are a complete wanker sometimes,' offered Simpson.

'I am,' I confirmed. 'Several times a day.' I smirked at Connor in a suitably self-assured way to get away with what I was about to say. 'Feel free to join in any time . . .'

Connor smirked back in a cheeky, flattered-but-not-yet, maybe, teasing sort of a way, then rolled his lovely green eyes.

'*Podex perfectus es!*' He looked deep into my soul as he said this, and the tone would have been ideally suited to the words 'I love you'. Sadly, the literal translation was: 'You are a complete anus.'

'*Podex perfectus habes!*' I replied tenderly, hoping both to retrieve the situation as well as display my grasp of Latin – 'you *have* a perfect arsehole'.

'You are *such* a poof,' he countered with a smile.

'You queer spastic!' This was Foster, enthusiastically joining the conversation. Two inches taller than me and built for a scrum. As such, I always treated him with more respect than he earned. After all, his preferred debating technique was physicality.

I had a full luncheon period, so I wanted to get on my way. I was soon through the huge double doors at the end of the main corridor, exiting the Great Hall (which always looked to me a bit like a red-brick cathedral) into blazing sunshine. In front of me, past the multiple tennis courts, I could see manicured playing fields stretching far into the distance. I knew that to my right, the other side of the long row of science labs, the playing fields stretched further until eventually they stopped at the swimming block. Behind me, on the other side of the Great Hall, was the exclusive expanse of the Common. I took for granted that, as befitted a member of the Eton Group, King's College School, in all its red-brick magnificence, quietly dominated the most expensive real estate of upper-middle-class suburban Wimbledon.

I'd never known any different. This was my world. I'd been educated at the primary school that fed King's since the age of three. At seven, I and most of my class simply crossed over to the other side of the Ridgway and started wearing a red blazer instead of grey. It was only now beginning to occur to me that I was privileged, and that (especially when my father had had to pay 95 per cent income tax under a

Labour government) I was only getting such a pre-eminent education because I was lucky enough to be born into the Establishment – as were all my extended family.

I'd always taken for granted that my relatives were well-off and important and connected. A high court judge here, several sirs and ladies there, a director general, a dean, numerous managing directors and chairmen. All of them claiming – in a rather reserved way – to love me; each of them within a few years of demonstrating the opposite.

As I walked on, Stinker strolled past, puffing on the pipe that had, unknowable decades before, dictated the affectionate nickname used for him by pupils of the junior school.

'Afternoon, H of S!'

'Sir!' It touched me that he still referred to me as head of school, even though that had been three years ago, when I was in the junior school. It helped strengthen my assumption that in six months I'd be made head boy of the senior school, meaning he wouldn't have to change his greeting. It was common for the same boy to be used twice. Certainly, within a few days I expected to be announced as a house prefect, so that must mean I was in the running to be head boy.

Through the windows of the Little Hall, I could see the full-width stage at one end of which, at thirteen, I had stood with the masters throughout every assembly, as head boy of the junior school. By the time I'd been chosen, it had become a largely ceremonial role, albeit (along with my band of prefects) with the power to frequently hand out conduct cards, of which too many led to a caning by the headmaster – although this was rarely administered to the seven-year-olds.

Things had changed. As some of the senior masters had relished telling me, one of my predecessors had had every one of his own prefects caned; by the time I took the role, this opportunity was evidently no longer viewed as a perk of the job.

Having skirted the rather large Little Hall, I was about to head to the Priory complex when Wiggy sailed into my path on his dilapidated bike and almost knocked into me, having taken the blind corner at full speed. I leaped to one side and waited for the crash.

Fortunately, he swerved to the other side; less fortunately, he almost slammed into the wall, inevitably hampered by having only one hand on the handlebar because the other was (as ever when cycling) glued to his head to hold his hairpiece in place. It was flapping in the breeze of his slipstream. This was better than a few weeks earlier when, during an unfortunate explosion in his chemistry lab, I'd seen it blow completely off.

'Careful!' he screamed, rather unfairly.

'Sir!'

I made it to the music block unscathed and, seeing a black vintage Rolls Royce parked under the archway, proceeded inside to find its owner. I entered a large room that was very familiar to me because as a member of the anthem choir (one of four choirs I sang in) I practised there before prayers for half an hour every day .

Mr Waters smiled warmly at me.

'Good afternoon, Peter. To what do I owe this unexpected pleasure?'

For any master to call a boy by his first name was unusual. However, in private Mr Waters always did.

'I wondered if I could leave my oboe here until tomorrow.'

'Of course, *mon petit*.'

After some brief chit-chat, I left and took the direct route along the whole length of the Priory, past the arts and crafts building and around to the art block. This was because, in addition to wearing a suit rather than a blazer, one of the main benefits of being in the sixth form was not having to attend luncheon. And in the upstairs art room, protected under a sheet, on its large reserved table, was what I'd been working on for the last eight months of lunch breaks. It was already, by far, my most precious and most intimate possession.

Pleasure

I pulled my sandwiches from my briefcase, prepared a quill and painstakingly added some calligraphy to annotate a minute image about a quarter of an inch tall, almost lost on the three-by-four-foot sheet of thick handmade paper. I pulled back to inspect the overall effect.

'It's beautiful!' a voice behind me broke in.

Larry Fish, one of the art masters, slim, in his forties, stylish, unmarried, I assumed gay.

'I didn't want to risk making a noise while you were writing. So, what *is* the "Flame of Analax"?'

Whether he was gay or not, I knew that Mr Fish was safe, so I told him the truth.

'It's where Lord Avalon first fell in love with Rahylan.'

'And Rahylan is you, if I remember right?'

'Sort of . . .'

It was complicated. Three years earlier I'd begun inventing another world, a fantasy world where the main action occurred in the Realms of Salania. I invented a geography, numerous cultures, a language and alphabet as well as cursive script and runes. At fourteen, I spent much of the summer holidays constructing and then carving a Salanian harp, which I still have proudly displayed in my office. Most importantly, I invented myths, sagas and ballads. And unlike Tolkien's works, which I'd loved, my stories of heroes and magic also had a major love story between two men. If the real world I saw around me lacked role models for passionate, dramatic and romantic love between two men, I would create my own to put that right.

This illuminated map – in the style of a fantastical medieval manuscript – was my record of every location, every story I had dreamed of. Every place name told a story. Every character had a history. Over the months, as I documented each one, Mr Fish would quiz me randomly on any intriguing wording that caught his eye. He knew that I identified with the blond apprentice warlock Rahylan, and I'd regaled him with unending tales about Rahylan and the Celt-like Avalon, but this was the first time I'd explicitly said that they were lovers. I told him of the ceremony at which the two declared their love to the whole court and then kissed in front of everybody, and how from then on they were formally welcomed throughout the realm as a married couple.

He treated the disclosure as being as natural as it was.

'You'll never follow the crowd,' he said, in a kind voice. 'I remember telling you that when you were nine years old and won the school art prize. I find it a shame you didn't choose art for A level.'

We'd discussed this a year before, when I'd had to choose which three A levels to study in the sixth form.

'And English, for heaven's sake,' he continued, back on a trail he'd beaten long ago. 'You'd be so happy in the art world. Or as a writer. *Or...*' he paused, as if the idea had just come to him, '... an actor. Mr Rodgers is still raving about that performance of yours as John de Stogumber. How you managed to break down in front of an audience every night...'

'All I did was cry in public. Even babies can do that.'

'Well, I expect he's going to make you the star of next year's play.' Then, in a sudden spurt: 'Or you could become a director, in film, television, something!' He looked exasperated, embarrassed and compassionate all at the same time. 'You'd *fit in* so very well.' His face contorted with the effort of trying to convey hidden meaning.

I assumed I knew what he meant but chose to continue as if all that we were discussing was my choice of academic subjects.

'I know. I love dram soc. And I'd have loved to do art and English A level. *And* geography and history. I've told you before, I feel I was born in the wrong era. Someone like Leonardo never had to choose science over art! I hated having to choose only maths, physics and chemistry. I couldn't even add biology – which rules out a medical career.'

'I just think that you might find more people who . . .' he slowed to choose his words carefully, '. . . *think* like you do, in the arts.'

'But it was always going to be science for me, always. By the time I was seven I was quoting Einstein's time dilation formula because I liked the way it sounded.'

'What?'

'It's the relativistic physics formula for how time slows the faster you travel.' He appeared no less confused. 'Look, if I took a rocket from Earth and went on a huge circular journey for a year, and kept accelerating for the first six months – only at the rate something falls to the ground – and then decelerated at the same rate for the return, then when I got back home, all my friends would be dead because on Earth a *hundred* years would have passed. It's like magic. But it's real. And that's what I love about science: just because you can explain something doesn't stop it being magic.'

'You see, that's exactly what I mean! You're going into science, Peter, but you're more at home with romance and love and magic, everything that's the complete *opposite* of science!'

'But that's the point – I don't think they *are* opposites, they're just different perspectives on the same things.'

A pause. He seemed to have acknowledged defeat and so came more directly to the heart of his well-meant warning: 'You do realize that you won't find many people who view the world like you do . . .'

I couldn't tell whether to respond to this as observation, accolade or criticism.

'Good! Anyway, I've devised a simple way to find who they are, if I ever meet them. I call it the Camelot Test.'

He raised his eyebrows in a silent 'go on'.

'So, you can be anyone in Camelot. Who do you want to be: Arthur or Lancelot?'

'I suppose most people end up choosing King Arthur?'

'Perhaps, but my answer is Merlin . . .'

He slowly smiled. 'OK, I think I get it. You know, I just wanted to double-check.' He looked at his watch. 'By the way, aren't you due for an inter-house match?'

I checked my own watch. 'Bloody hell!'

'Don't worry, off you go.'

I snatched up my briefcase and ran enthusiastically towards what, almost half a century later, remains the most traumatic episode of my life.

Once a year, the six houses of the school held a fencing tournament. It was about to start and I was competing. Unfortunately, I was also still wearing a suit.

I charged into the almost deserted changing room of the fencing salon, grabbed my fencing bag from the rack and began to undress.

'Fashionably late, I see.' Waspish as ever, Anthony wandered over, smiling a welcome.

'Anything to throw Nicholson off his game!'

Anthony: heavyset, in the year above me, in a different house, Hungarian parents (his mother had been in Auschwitz), not interested in science, planning a career in law, not really bothered about fencing, not one but two girlfriends, and my best friend. Somehow we just got on. And he was a vicious mimic.

'Scott!' This was his highly exaggerated version of my housemaster, whom we both despised. 'Late again! The shame, the shame, to the house, to the school, to the world!'

We both dissolved into gales of laughter. By now I was fastening the last shoulder buttons of my white fencing jacket.

'Am I all set?'

Anthony quickly inspected my kit.

'A vision in white,' he concluded. 'Good luck!'

'I'll never beat Nicholson, but I'll try not to make a complete prat of myself . . .'

I grabbed my mask and foil and ran into the salon.

Almost two hours later, I found myself in the final, matched against Nicholson, a year older than me and head of fencing. He was definitely the better fencer, but the maître d'armes had avoided the two of us fighting in earlier play-offs – which was largely the reason I'd got so far.

As it turned out, I was having a good day and Nicholson was off form. We were neck and neck, and the first of us to get another point would win. I was waiting at the end of the piste, breathing heavily, with my mask off, to cool down between bouts. I was trying to think. Every trick I'd tried he'd got smart to. He was quick and clever and experienced. I felt sure he was going to get the next point.

I took a deep breath and forced myself to calm down. Think! At its heart, fencing is just a very high-speed chess game. To score, you simply have to get your opponent's sword out of the way long enough to strike. It's called a parry-riposte. Really, there are only two ways to parry: you bang your opponent's blade to one side or you twirl around it and flick it away. Trouble is, your opponent knows this and tries to counter everything you do. And Nicholson was very fast. So how could I outthink him?

And then I realized the pattern to how he fenced. When we started a bout, Nicholson's play seemed quite random. But when he felt under pressure, and resorted to instinct, he seemed to favour circular parries. No time. Mask on.

'En garde! Allez!'

Immediately, I attacked hard with a flurry of parries and ripostes. I threw all my energy into a burst of movement I knew I couldn't keep up for more than about thirty seconds. This was all or nothing.

I stepped forward aggressively, forcing Nicholson down the piste. And then it happened. He executed two circular parries in a row. I gambled he was about to do a third, avoided it in advance, banged his blade out of the way and lunged.

He leaped back, just out of reach of the tip of my foil. But I had long legs and had lunged so fast that my momentum carried me forward. I pivoted on my right leg to gain an extra few inches.

'Touché!' The maître d'armes sounded as excited as I was. I whipped off my mask at just the same time as Nicholson and we both shook hands. He was smiling his congratulations and I was vaguely aware of everyone in the salon cheering and surging towards us.

'Good show, Scott! You'll make a worthy successor.' Nicholson seemed completely genuine.

I felt elated and immensely proud. After I'd reached the quarter-final of the Public Schools Fencing Championship,

I'd been told I was in line to take over from Nicholson this term so he could concentrate on the A-level exams that began in a month's time. But this was him unofficially handing over the mantle.

My housemaster – large, white-haired, intimidating – pushed his way through the crowd, no doubt to offer his grudging congratulations. But instead, he avoided eye contact and signalled to the maître d'armes. They walked away and huddled together in a corner whilst everyone who had been watching the final huddled around Nicholson and me. I didn't even notice my housemaster's return.

'Scott! Come with me.'

I followed him out of the salon into the changing room.

'Leave them here,' he ordered, waving at the foil and mask under my left arm, then stormed out without even checking that I was in tow.

We marched in silence and had already reached the tennis courts when he answered my unasked question.

'You are coming with me to the headmaster!'

This was a euphemism. Masters of all levels of seniority, but especially housemasters, valued their autonomy. They maintained extensive powers of corporal punishment that some, most notably the older teachers, continued to exercise with the help of a hard-soled slipper or old gym shoe. But these days, only the headmaster was allowed to cane. Indeed, to most of us boys, caning appeared to be his *only* role other than appearing during morning assembly.

My stride never faltered but I suddenly felt sick. My mouth dropped open a bit.

'Sir, may I ask why?'

'It's a disciplinary issue,' he barked.

The Disabled

All my esoteric blood results came back normal. On the plus side, I had now been tested for almost everything for which there are blood tests and I had passed with flying colours. On the minus side, there were no more blood tests left, yet I still had *something* wrong with me. Something that was by now far more debilitating than just badly wiggling toes; climbing the steps of a Mayan pyramid temple in Mexico was becoming a health and safety issue.

Six months after my first symptoms appeared, the partial paralysis of my right foot had spread up to my knee. And it was now bilateral – in other words, everything that had started on my right leg had gradually been mirrored in the same sequence on my left. As Francis and I explored a Norwegian fjord one day, I realized I was walking with a very obvious limp.

Back in the UK, none of my growing medical team yet had the slightest idea what was causing this – although, on the positive side, we had a prodigiously long list of all the things that it wasn't. Perhaps spurred on by their increasingly embarrassing failure at diagnosis, they did the next best thing: they tacitly agreed to give my symptoms a suitably impressive name. Henceforth, I was officially suffering from 'spastic paraparesis' – that is, stiff leg muscles causing partial paralysis.

This prompted me into a re-evaluation of my old classmate Foster and his gang. I distinctly remember that he and a small but voluble subset of the school chose to reference my existence with the descriptive epithet 'queer spastic'. At the time, I'd put this down as puerile banter. But seriously, to end up being absolutely correct on *both* counts . . . What are the odds?

By this stage, when we were out, Francis was increasingly offering me his arm if it looked like I needed assistance. It was, I justified, the same subconscious gesture that any couple who'd been together all their adult lives would tend to display. Sure enough, I often saw little old ladies shuffling along holding on to their husbands in just the same way as I clung to mine. Just as Ma and Da had done.

Then again, they'd been in their nineties at the time. I was in my fifties. And I had a disconcertingly physically fit and able-bodied look about me. It slowly dawned on me (not least when I registered the exaggeratedly caring or mildly repulsed looks on strangers' faces) that passers-by were not pigeonholing us in the same way as they did the old dears making their way to the benches along the promenade in Torquay. They didn't even categorize us as a couple of old queens. In fact, they didn't think of us as a couple at all. Instead, Francis now seemed to all intents and purposes to be my personal care assistant. And, in a mere handful of months, I had unknowingly metamorphosed into what appeared to be an adult with learning difficulties. This was my very first experience of joining my new tribe: the disabled.

I must have been in denial for at least four months before I finally admitted to myself that maybe this was not just a phase I was going through. Maybe I wouldn't just experiment with the edgy lifestyle of being disabled and then return to the more acceptable standards of the mobile majority. Maybe this was who I was from now on. I plucked up courage and told my best friend (Francis, of course) that I thought I might be disabled. As in, *really* disabled. He said he already knew. Which was a huge relief. But also, a big deal. I remember us bolstering our resolve to once again face head-on any ignorance and prejudice that came our way, with the words: 'Thank goodness it's only the legs, eh?'

As it turned out, the big difference between coming out as disabled compared with my first experience of coming out as gay was that this time people tended to be really nice about it. In which case, I reasoned, what's not to like? This extraordinarily altruistic demonstration of common humanity was best demonstrated on the Mediterranean island of Ibiza.

Francis and I found ourselves crossing a busy road. Slowly. Me clinging on to his arm as the traffic lights – timed for vigorous young partygoers – beeped aggressively in a vain attempt to hurry us up. By the time we were halfway across, we had been left far behind by the hordes that began their migration at the same time as us. Our only remaining travelling companion was a wizened old woman in black widow's weeds, apparently intent on nothing other than getting across before she could be mown down by a surge of traffic. She was in the lead.

Then the lights changed and I heard the revving of engines and the mosquito duet of a couple of overstrained scooters racing each other away from the still-amber lights. Towards me. Brilliant.

Now, at this moment, you would be forgiven for assuming that our companion road-crosser (well accustomed as she no doubt was to the unrelenting hostility of the local traffic) would have put on a tortoise-like burst of speed, following the logic that she only needed to be faster than the couple behind her for the scooters to hit us first and screech to a suitable stop.

Instead, as the rumbling whine of traffic got louder, she paused, turned back and took my other arm. Francis and she then escorted me to safety like a couple of slow-motion security guards manhandling an unwanted visitor off the premises. After she'd delivered me safely to the kerb, she smiled toothlessly at Francis, gabbled something at him in Spanish, turned and hobbled away. She never made eye contact with me. This was a watershed moment for me. I realized that I had just unwittingly and unwillingly transitioned two momentous rites of passage. Firstly, after a lifetime of at least being tolerated because of my brain, I was now perceived as an idiot, not worthy of even being acknowledged. Secondly, I had reached a stage in my life when little old ladies were helping *me* across the street.

Something had to change, and intuitively I sought a technological solution. I quickly concluded that the pinnacle of walking-stick technology had been reached in the late-Victorian age in the form of the gentleman's dress cane. This elegant but sophisticated design was never surpassed, in my view – a point that was rammed home to me when, back in England, I asked my physio to measure me for a walking aid and he lent me the stick that was available on the NHS.

It was basically a wide-diameter adjustable aluminium tube with a hard, straight, grey plastic handle on top and a huge rubber ferrule on the bottom. It was surprisingly heavy, badly balanced, and the handle was uncomfortable if I put any weight on it. The whole ensemble looked like it had been designed by a moonlighting plumber.

It rattled. It clicked when I walked with it. But mainly it screamed *sad*. I could only assume that whoever had authorized the design had never actually had to use one in public. In contrast, the perfectly balanced, antique, silver-topped ebony cane I'd just bought had been designed in the era of Sherlock Holmes to scream *debonair sophistication*. Even 120 years later, it still won effortlessly against its modern clinical

26

counterpart, which (forget about looks) didn't even work properly.

When I first strolled out with my new/old walking aid – a hugely symbolic moment for me – I think I actually had a bit of a swagger about me. Then again, it might just have been my limp. Either way, no one ever again mistook me for having anything other than an obvious physical disability. Indeed, passers-by would often briefly look me in the eye. And smile. It's amazing what displaying the correct membership badge can achieve . . .

Pain

I don't even remember how we reached the antechamber outside the headmaster's study. My only recollection is my housemaster bursting through the swing door and scowling to discover that the small room was already occupied by two boys, heads down.

I knew one of them reasonably well. Perched on the bench against the wall was Rawlings from my year, wearing red-and-blue rugger kit and mud-caked boots, just off the field. Sitting beside him was my lovely friend Connor, in white gym kit, back from a cross-country run. For a moment they both looked up, surprised at the intrusion. But after a few seconds of meaningful eye contact with me they looked down again, trying to be strong.

Rawlings, whose height and build made him appear two years older than he was, looked stoic; little Connor, shorter and lighter than me, looked close to tears. My heart went out to him. Meanwhile, my housemaster stood as if there was no one else in the room. Through the padded door to the headmaster's study I could hear the muffled drone of a one-sided conversation.

Suddenly, the door opened and I saw Bellchamber, much the same physique as Rawlings and also in rugger kit, but with hairy legs and tear-filled eyes. The headmaster stood behind him, old, slim, tall, thinning grey hair, slit eyes, tanned reptilian skin.

'Next!'

Eyes fixed straight ahead, Rawlings stood, and the

headmaster pushed Bellchamber on the shoulder to get him moving. He tentatively hobbled forwards, his legs locked as if he were unable to flex his knees. He gave Rawlings an almost imperceptible shake of his head. Not a good sign. The thick navy material of his rugger shorts had evidently been considered an unfair advantage and he'd been made to pull them down.

The door closed on Rawlings. Dull drone. Silence. My heart was hammering. My face felt cold.

Shhwukkk!

Silence.

Shhwukkk!

Silence. I was suddenly hot. I felt close to throwing up.

Shhwukkk! and a simultaneous shout of pain.

Silence. A whimpering sound to my left. I looked across. Sweet darling Connor looked terrified. I wanted to hug him. All my instinct was to protect him. He looked so slight and vulnerable. I'd had a crush on him for the last year, but for now all I felt was his abject misery – because I too was terrified.

I'd never been caned before, but we all knew exactly how our friends described it. You were told to bend over the chair and look out the window. Then you waited. Everyone said the waiting was the worst. Then the first stroke. Immediately painful but after two seconds excruciating. Then the second stroke. That's when, however strong you were, you started crying. Then the third stroke. Usually after a longer delay. That's when most boys made a sound. And then you were supposed to stand up, hold your arm out, shake hands with the headmaster and say, 'Thank you, Sir.'

The door opened. 'Next!'

Connor stood on his pale slim legs. And didn't move. He was petrified. Rawlings stiffly emerged, his face bright red but with strange white patches, drained of blood, around his nose. The studs on his rugger boots exaggerated the staccato sound of his slow bird-like walk out of the antechamber. Connor didn't move.

'Get on with it, boy!' My housemaster, more exasperated than ever. The headmaster frowning at Connor.

That did it.

My terror remained. But for the first time in my life I felt stronger emotions building on top of it.

An outrage against reality. Anger at injustice. Hate of a cruel system.

An alien calm numbed my panic. Despite my impotence – my utter powerlessness to save little Connor, or myself, from the inevitable – I was increasingly infused with a sense of resolve, of responsibility, of willing sacrifice. Of power.

It was such an unfamiliar cocktail of emotions that I didn't know what to do with them. I felt grateful to no longer only feel scared. But, as Connor shuffled into the study and the door shut, I realized that, if such a trade had been conceivable, I would have willingly taken his place. I cared for him. I was stronger than him. And a wakening part of my psyche was screaming that with strength comes responsibility.

Yet all the rest of me was screaming: *You can do nothing.*

With the muffled drone, I felt my teeth clench.

With the silence, my face froze.

Shhwukkk! A cry.

Outrage.

Shhwukkk! A louder cry. My housemaster casually checked his watch.

Anger.

Anger.

Anger.

Shhwukkk! A short scream.

Hate.

I was seething. Outrage. Anger. Hate. At least it was now ov–

Shhwukkk! An urgent scream.

Oh, Christ! No! How could he?

Shhwukkk! A howl that continued as a sobbing wail.

I *hate* you!!!

Wailing.

I HATE you with every fibre of my being.

Shhwukkk! A high-pitched scream followed by the equally unfamiliar sound of a sixteen-year-old boy crying uncontrollably, jerking between tenor and treble as his vocal cords randomly reverted to the last time he'd broken down as a child.

Please, God, please, no more! In my desperation, I pleaded with a divine being that I knew didn't exist.

Please. Please.

Muffled murmuring.

Did that mean his suffering was over? I couldn't hear him crying any more.

Silence.

Louder murmuring.

The door opened and there was beautiful Connor. Not so beautiful now, the sides of his mouth pulled down into an impossible semicircle, his green eyes bloodshot, and his pale cheeks streaming with tears as he extended his humiliation by continuing to cry in front of us.

All my instinct was to hold him tight and kiss away the tears and tell him it was over and I'd never let it happen again and I would protect him.

Instead, for a few seconds his eyes met mine and I tried to convey it all with the hint of a compassionate smile and a very slight nod. He acknowledged me with an almost subliminal nod back, looked away and tried to walk out.

He took pigeon steps towards freedom. As he reached the swing door and, now with his back to me, paused to push it open, I saw a faint red stain developing on one side of his shorts.

I expect my face showed nothing, as my brain was far too busy recalibrating to find time to control muscles. My headmaster had always been the unquestioned leader of my sequestered existence. I now saw him to be the despotic tyrant of a brutal regime of terror. All my life I'd been groomed, brainwashed, to accept this Establishment as good, worthy, appropriate. Suddenly, its unopposed institutional bullying and cruelty utterly repelled me.

I knew perfectly well that I could do absolutely nothing to change what was about to happen. But, like Rahylan and Avalon going into battle, I would not give them the satisfaction of capitulating. I would not let them sense the fear that I felt. I would not give in. I would be strong. For Connor.

I raised my chin and turned to look the headmaster straight in the eye.

'Next!'

Shocking Developments

The continued absence of any diagnosis was now getting tedious. So, to my boyish delight, we began a series of far more obviously hi-tech tests involving proper electronic equipment with flashing lights, electrodes and computer screens – the way Hollywood would do it.

The first experiment was almost nothing *but* flashing lights. It was a visual evoked potentials (VEP) test to check the nerve messages between my eyes and brain. It made sense to do because it is one of the common tests for MS, which as I'd long suspected was now a strong contender for being My Disease.

A technician attached some impressive-looking electrodes to my skull, covered one of my eyes and left me alone in a quiet room, and I sat watching some flashing patterns until he returned and swapped eyes. It was remarkably like how I imagined brainwashing experiments must have been during the Soviet era.

Next came a far more impressive-sounding somatosensory evoked potentials (SSEP) test to check the nerve messages between my body and brain. This, I quickly discovered, was clearly derived from *interrogation* techniques refined during the Soviet era. The giveaway should have been the electrodes attached to various strategic parts of my body.

Then the electric shocks started. About two a second. Sharp but bearable.

'How's that?' the doctor asked. She was quite young, still displaying the enthusiasm of the recently qualified.

'Fine, thank you.' You hear varying reports on SSEPs, but this was going to be OK.

'Oh, good! I like to start at about half the voltage just to check. Though you're slightly taller than average so we may have to go a little higher with you. Are you happy for me to ramp things up a bit?'

She was infectiously keen, so I found myself lying through my clenched teeth: 'Yes, absolutely.'

The thing about electric shocks is that it's the amps that kill you but it's the volts that make your muscles contract. As my eager young companion turned up the voltage, I started twitching. And everything became progressively more uncomfortable. She hadn't even reached max yet.

'Now, I need you to relax for me,' she chirped.

I studied her for any hint of jocularity and reluctantly concluded she was serious.

'OK, let me get this straight. I'm wired to a gurney, you're about to whack the voltage up to full (or maybe more because I'm taller), and you want me to relax?'

'Yes, that's perfect. In fact, if you don't mind, in order to get the best possible results, I'd like to push the voltage as high as we can while you're still able to keep your muscles relaxed. Don't tense them. Just let them twitch. That would be great, if you're up for it?'

She'd said the magic words 'best possible results' so Peter the Scientist naturally replied: 'Of course. You go for it!'

Just as the voltage was moving from uncomfortable to memorable, I thought to enquire a little more about the test.

'How many electric pulses' – for some reason that sounded less worrying than 'shocks' – 'are needed for the test?'

I tried to sound casually intrigued; I was actually very interested in the answer.

'Oh, I think we'll do six hundred.' She seemed pleased with this answer. 'Then we'll change sites.'

'Ah, great. And how many sites do you plan to cover?'

'Four,' she said, as if clarifying the number of walls to her torture chamber.

With the 'pulses' now elevated from memorable to unforgettable, I sought consolation by asking for a progress report. After all, I must have been being shocked for at least five minutes by now.

'So, where are we regarding the first site?'

She beamed across at me from her computer screens, knobs, buttons and other instruments of duress.

'Don't worry. Another minute – and then we'll be ready to start . . .'

A fortnight after I'd been released from my SSEP, I'd received official confirmation that despite the doctor's best efforts I'd told her nothing. I already knew this. I'd kept talking to my interrogator throughout – albeit in a sometimes slightly squeaky voice during the more unpleasant bits – asking about her results. They'd all been normal (just as, I'd learned, had the VEP results). When she'd finally admitted defeat and said I was free to go, she'd seemed positively deflated.

I should have guessed the authorities wouldn't leave it there. Only a week later, I was hauled in again. This time, they told me that they accepted I didn't have MS. They were looking for something else. When I asked what, they said they weren't sure. They wanted to 'rule out motor neurone disease once and for all'. I said I thought the MRI had long ago ruled that out. They agreed. But anyway, they were going to link me up to an EMG to see if there was something I was hiding.

The electromyograph is a diagnostic tool for neuromuscular diseases, but I suspect it would be far more at home in a James Bond movie. Not in Q's lab, I hasten to add – in the villain's stately-home lair. You know the trope: an incongruous battery of hi-tech equipment and white hospital bed jarring against the high-ceilinged wooden-panelled décor of a gentleman's study commandeered for the mandatory torture scene. Inevitably, it would be operated by a disconcertingly quiet middle-aged woman, devoid of make-up and dressed in slightly old-fashioned clothes.

By an extraordinary coincidence, that's exactly how I now found myself, with the lair masquerading as a rural cottage hospital outside Taunton and my suburban tormentor concealed under a consultant's white coat. She talked quietly at me in an Eastern European accent I couldn't quite place.

On the surface, from the prisoner's point of view, there is a shocking similarity between the SSEP and the EMG. The difference, it turns out, is *how* those shocks are administered. The EMG is a back-to-basics type of interrogation. It maintains the strategically placed first electrode (taped to a suitably exposed segment of your flesh) but utterly repudiates any namby-pamby electrodes to the brain in favour of an electrocuted needle – that is slowly pushed through your skin into various muscles in turn. And then wiggled. As the pulsed electric shocks continue. Unrelentingly.

No doubt tipped off by her SSEP colleague, this particular Angel of Pain refused to be drawn into conversation throughout the whole of my harrowing inquisition. She'd stare intently at her screen, move the needle in and out of my flesh, then leave it undisturbed while it continued to shock me. Eventually, apparently satisfied by something, she'd draw the needle out, gently caress another muscle and slowly push the needle in again. When she got bored with my feet and legs, she moved to my hands and arms. I didn't give her the satisfaction of reminding her my arms were fine – I sensed she'd go ahead anyway.

I was particularly interested in the test results when they came through. It was a rather technical five-page letter addressed to my neurologist that I'd requested I be copied in on. By page three, I realized that the conclusion was, once again, that my test results were normal. In a nutshell, there was no breakdown in the insulation around my lower motor neurones (the nerves going from my spine to my muscles). So, that ruled out motor neurone disease. Just as the MRI had done.

I was starting to lose interest in the letter as it proceeded to confirm all the various diseases that the EMG results now ruled out. Just as I'd thought, there in the list was amyotrophic lateral sclerosis (ALS), the technical term for MND. More precisely, I vaguely remembered, ALS was a *type* of MND – the nasty type, by far the most common one, that in the USA was called Lou Gehrig's disease. As far as I could tell, everyone in the medical profession that I'd met so far wasn't too bothered by this pedantic distinction and instead tended to use 'MND' and 'ALS' interchangeably.

So, I didn't have ALS/MND. No surprise there. I skimmed the rest. But then something arrested me in my tracks. What was this? Suddenly the letter had my full attention. Here was a tentative proposal for a candidate diagnosis. A brand-new disease that no one had mentioned to me before. An arcane grouping of words that I was pretty sure I had never ever even heard before: primary lateral sclerosis, PLS.

My Turn

I strode into the headmaster's study as I hoped a teenage prince going to his execution would stride. Fearlessly. Inside, I was trying to suppress my primal fear with a white fury, two untamed forces held in balance through strength of will. Just. I was young and inexperienced in how to maintain equilibrium, but the arrogance of youth gave me the self-belief I needed – I had faith in my willpower.

My housemaster slid in behind me and closed the door as the headmaster, with slightly red cheeks I noticed, walked to his desk. And there it was. Almost three feet long. Thick as my finger. The one reserved for the older boys.

He turned towards me, his back to the large sash window that looked out on to the idyllic public school that was my life. He stared at me.

'Sir!'

My housemaster intervened from behind me.

'You'll remember, Headmaster . . .'

'Yes, yes, very unfortunate business, Scott. Distasteful. Most disappointing.'

'Sir?'

He appeared to be contemplating how best to proceed. His unblinking eyes were those of a snake waiting to strike. When it came, his voice was double its normal volume.

'Do you *want* to be an abomination, Scott?'

Fuck!

It was the last thing I'd expected. I was completely unprepared. Nobody referred to someone as an abomination unless they were making a deliberate reference to Leviticus.

'Well?'

My mind was blank. Of all the reasons to be here, this was the worst. Literally the most damning. I replied on autopilot.

'Absolutely not, Sir!'

He was shaking his head in either disbelief or disdain.

'Well, that's not what I hear, Scott. Not what your house-master hears.'

Taking this as an invitation to contribute, my housemaster joined the headmaster at the desk – the cane symbolically in view between them.

'He's developed a highly inappropriate reputation, Head-master. Not just in his year – the younger boarders openly joke about it over dinner. Very, very distasteful. He sets a totally unacceptable example to the whole school, Headmaster!'

My brain was frantically decoding the patterns in what he'd said. At least I now knew the most likely source of his information. King's was primarily a day school, but for a small minority of boys it was a boarding school. Connor was one of them. Of course he'd joke about me, just as he did when we were joking together. Of course he'd gossip; I loved it when he gossiped. Of course other boys would hear.

In this case, it seemed, a master had too.

It wasn't Connor's fault.

'You do realize, don't you, Scott, that being a *sodomite*' – he stressed every syllable – 'a *catamite*, a *queer*' – he managed to insert two syllables into queer – 'is an abomination against God and humanity?'

My eyes flicked to the cane and imperceptibly I took a deep breath to calm my thinking. This was a tricky question to answer. 'Lying with another man' was only ever mentioned twice in the Bible, both times in the Old Testament, and both times by the rather intolerant Leviticus – who also called shellfish and tattoos abominations, and who insisted that non-virgin brides and disrespectful teens should be stoned to death.

'I believe that's only in Leviticus, Sir.'

'It has nothing to do with *where* in the Bible it says it, boy! It is against all common decency. It's a disgusting perversion of God's natural order!' He paused. 'You *do* believe in God, I take it, Scott?'

Another tricky question.

'I'm afraid . . . I'm an agnostic, Sir.'

It was an outright lie. Agnostics aren't sure whether there's a God. By fifteen, I'd felt certain she was manmade.

'Aah!' He turned knowingly to my housemaster. 'This explains why he's lost his moral compass.'

Turning back to me: 'In which case, at least you will accept that it is *illegal*, I assume!'

Sadly, this was true. I was painfully aware that six years earlier, when I was just old enough to vaguely understand but just too young to realize its relevance to me, homosexuality had been decriminalized – but only for adults, in private, behind locked doors. My twenty-first birthday was a lifetime away.

'Yes, Sir.'

'Now, you may have heard that in special circumstances these perverts are no longer sent to prison. But you do realize, don't you, that just because something is legal does not make it acceptable?'

He was a magistrate so clearly felt he was an expert on interpreting the law. I entered into the spirit of our Socratic dialogue and answered truthfully.

'I did *not* realize that, no, Sir.'

This appeared to placate him.

'Look, Scott, I understand that you are possibly confused. It is a perfectly natural phase to go through to feel strong friendships with other boys. But trust me when I tell you that inevitably this will pass. When you are eighteen and have left school, you will find a girl and get married, and she will have your children, look after the house and be waiting for you when you come home, and she will be able to satisfy your natural manly urges. It's the order of things. It's what real men do.'

It was the longest speech on manhood that I'd ever heard him give. I found every phrase of it repellent – which made me all the stronger. I looked blankly at him. This only seemed to encourage him.

'In contrast, ho-mo-sex-u-al-it-y is a disease. It's the worst sort of *venereal* disease. It leads to a guaranteed life of wretched misery. Ho-mo-sex-u-als are universally sad, don't you know, Scott.'

He paused to allow me time to ponder this Great Truth. I used the opportunity to reflect on just how much I loathed him, his views and the whole Establishment that promoted people like him to positions of power.

'I did not realize that either, Sir, no, Sir.'

'Well, that's that then.' He didn't sound as if he'd finished, merely completed the first order of business. 'I'm glad we're of the same mind. You see, it's all about life choices. It's good to have had this little chat before I discipline you.'

Heart suddenly thumping. Stay in control. You knew this was coming. You just forgot. Be strong. Hate him. Remember Connor.

'This is going to be a harsh punishment, Scott. I want it to last. I want you to remember it whenever you're presented with life choices. I want it to be a painful memory of the consequences of making the *wrong* choices. From now on, I expect you to live a life of moral rectitude. Although I know this will hurt, for the rest of your life you will remember today as the day you pulled your socks up and began a new life. Do you understand?'

Be brave. How would Rahylan do it?

'Yes, Sir. I understand completely.'

He turned towards the cane on his desk. My housemaster moved to one side.

Outrage. Anger. Hate. Connor.

But it's so unfair.

Be brave.

'This letter, Scott . . .' He was facing me again, with a sheet of foolscap in his hand. 'It's from the maître d'armes proposing you be made the head of fencing. Clearly, this will not now happen. It would send quite the wrong signal, to have you representing the school and so forth.'

And there it was. In just a few words, he exploded my world with as much compassion as he beat boys. Fencing was my passion. It was the one sport I was good at. He knew that.

Be brave.

'Also, I want you to leave the dramatic society immediately. It's an inappropriate direction of travel.'

A second hit, straight to my core.

Even through my shell shock, my brain suddenly realized what he was doing. I could decode the pattern. He was pausing. He always paused before the third stroke. It was that one that inflicted the most pain and made you cry out.

Be brave.

Silence.

Be brave.

'Finally, your housemaster and I are in full agreement that, under the circumstances, we cannot possibly make you a prefect at any time. It's a crying shame. I saw you as a serious contender for head boy next year. Huge disappointment. For your parents as well.'

I was numb. In only a few seconds he had casually destroyed everything he knew was important to me. Everything that made me feel special. Everything I was proud of.

He held his arm out.

'That's all.'

My caning was over.

I shook his hand.

'Thank you, Sir.'

Self-discovery

I took stock of my situation. After eight months of slowly progressing paralysis of my legs, pretty well every possible cause had been ruled out. The MRI of my brain and spine was clear. An increasingly comprehensive and esoteric raft of blood tests were all clear. My genetic tests were clear. My VEP and SSEP were both clear. My EMG (wonderful form of legally sanctioned enhanced interrogation as it was) was clear.

By a triumph of elimination, this left primary lateral sclerosis (PLS) – a rare, slowly developing and largely benign form of MND. Amazing. Also fascinating. This was a working diagnosis I could live with. Literally. And the fact that it was rare and exotic and hardly anything was known about it was a distinct bonus. It felt like the ultimate reprieve. I was off the hook.

And I was back in my element. Habitual scientist as I am, I began to study PLS intensively. It was a bit like undertaking the literature survey at the start of my PhD research all over again. Back then, I'd been excited by the creative potential of the challenge and had transformed my survey into a book, *The Robotics Revolution*, in which I laid out my thinking on the future of AI for the first time. But now there was a sense of urgency about it. Fortunately, there wasn't that much to learn.

In essence, it turned out that PLS was a disease of only the *upper* motor neurones, whereas ALS (the far more common, albeit fatal form of MND) destroys both upper *and* lower motor neurones. Easy translation: *neurone* is simply another word for 'nerve cell'; the *motor neurones* are long and thin (just like the electrical wires they mimic) and go all the way from the brain to each of the muscles that you can move just by thinking about them (even the medical profession uses the word 'motor' pretty much like roboticists do); each *lower motor neurone* runs from its particular muscle all the way to the spine where, a bit like in an electrical junction box, it connects to a corresponding second wire – called an *upper motor neurone* – that runs up the spinal cord and into the part of the brain that controls voluntary movement.

So, in everyday language, some of the wiring coming straight out of my brain was slowly breaking down. In the scheme of things, that was damage I could cope with; it's the loss of the *lower* motor neurones that causes muscle wasting that, without intervention, kills you because you can't eat or breathe. PLS, in contrast, may even leave you able to walk. It was brilliant news. Francis and I were in jubilant mood – like we'd both unknowingly dodged a poison bullet.

After three months' study, I felt I knew as much about PLS as many of the consultants I met and also felt I could hold my own in a peer-level discussion with them. But there was a problem: I was confused. It wasn't the kind of frustrating confusion that comes when you don't understand something. It was the strangely alluring kind of confusion that comes when you *do* understand something sufficiently well that you begin to sense that it just doesn't add up.

It was a feeling I'd become very attuned to over the years. It was what got me into science in the first place. And I love it. It means that you may be on to the trail of something new and therefore exciting. If you're correct, and if people agree with you, then you get to rewrite the future. It's an amazing feeling. Why else do you think geeks spend their lives in labs? So, this sort of confusion I was normally intuitively drawn to resolve. But not in this case.

The trouble was that it wasn't PLS that didn't fully add up, it was ALS (the common form of MND). And that wasn't the field I was interested in. As it was, I'd only ever unintentionally learned about ALS because PLS was *so* rare. There aren't really any scientific papers exclusively about PLS. Instead, it's a condition that – on the infrequent occasions that it's mentioned at all – is mentioned *as part of* some more general MND research. Usually in the context of ALS.

Long story short, it's impossible to learn much at all about PLS without as a by-product learning a *lot* about ALS. And so it was in my case. Eleven months after my first symptoms, I'd memorized what little I'd been able to uncover about PLS – but I was a walking encyclopaedia on ALS. That's how I knew that what people said about the science of MND and what the science *itself* said often didn't match. In some really crucial areas.

For instance, MND was always described as rare. But when you burrowed down into the epidemiology statistics, the chances of any one of us getting it in our lifetime was actually 1 in 300. That didn't feel especially rare to me. I doubted that many parents or teachers realized that roughly three of the kids being taught in an average UK secondary school on a given day would eventually die of MND. In fact, put like that, it didn't feel rare at all.

Far more confusing to me were the mortality statistics: 30 per cent dead within one year, 50 per cent dead within two years, 90 per cent dead in five years. How could this possibly be true? After all, despite the universal claim that MND was 'invariably fatal', this wasn't strictly true; despite the equally universal claim that MND was 'untreatable', this wasn't strictly true either.

My point was that people with MND – technically, those with ALS, the common form of MND that swamped the statistics – died from starvation (because they could no longer swallow food) or asphyxiation (because they could no longer breathe). But why did this happen?

Their digestive tracts remained fully functional, so they could easily be sustained using the very common technique of a feeding tube directly into their stomach. Their lungs still worked, only the muscles needed to inflate them got weak, so they could easily use an air-pump to inflate their lungs instead. The causes of death were not really medical issues – they were more like engineering issues.

It seemed to me that, treated with the right technology, MND was far closer to a chronic disease than a terminal disease. Use the right technology and it might not be MND that eventually killed you – it might be heart disease or cancer.

So why, despite this, did so many die so soon? It was as if most people with MND, for whatever reason, simply didn't take the technology route in order to stay alive. Maybe they didn't know about it? Maybe they weren't offered it for care-cost reasons? Maybe they didn't *want* to keep going?

I could at least understand this last one. Historically, MND was a truly vile disease. If you used life support to stay alive then you eventually became almost totally locked in, able only to move your eyes to explore from your bed what was typically a rather boring hospital ceiling. But thanks to emerging hi-tech, that had all changed, of course. The latest developments made even someone like Stephen Hawking look lo-tech. Everybody knew that things were radically transforming. Didn't they?

I didn't waste any more time on it. Intriguing though this avenue of enquiry was, it was irrelevant to my PLS focus. Nevertheless, it *was* intriguing. In the absence of anyone else with whom to share what (to me at least) were fascinating but apparently immaterial insights, one evening over a glass of red wine I explained in detail to Francis that – despite the overwhelming weight of fake news concerning prognosis – the truth was that if you wanted to live with MND then the odds were very much in your favour that you could. He agreed that it was curious that this truth wasn't made clearer. And, more importantly, why wasn't there more information about PLS?

It was in this state of mind that, three weeks later, Francis and I travelled the 200 miles to London for a couple of days of repeat tests. It was approaching a year since my very first symptom, and our trip was the ultimate example of dotting every 'i' and crossing every 't'. None of my tests had revealed anything, so I'd asked my GP to let me take advantage of an extraordinary service offered by the NHS that a lot of people didn't seem to even know about – namely, anyone can request to be treated anywhere in the country.

I opted to go to the National Hospital for Neurology and Neurosurgery. Where else? People came all the way from foreign countries to be diagnosed there; I only had to get up from another county. Well worth it to confirm the PLS diagnosis. They'd repeat all the most important tests. Oh, and they'd give me a lumbar puncture (also known as a spinal tap) for good measure.

I was lying on my left side with my legs drawn up in the foetal position. It was a primeval position of total submission, my main organs protected as well as they could be against the physical abuse about to be administered by the man towering over me.

The doctor had in his hands a very large syringe with a disconcertingly long and thick needle attached. I knew this because he'd proudly displayed it to me a minute before – just after I'd signed a waiver form that basically said I wouldn't sue him if he permanently paralysed me with the procedure he was about to perform.

As far as I can tell, lumbar punctures have a bad rap. Just because a total stranger (who, if you're honest, you can't be certain isn't just a psychopath off the streets who's stolen a white coat and walked into the treatment room), out of sight behind you, shoves the equivalent of a knitting needle into the small of your back and then lets your watery cerebro-spinal fluid slowly drip out into a small vial, this seems to give people the impression that it's something nasty and to be avoided. Certainly, the way Hollywood always depicts it you'd have thought it was the stuff of phobias and nightmares and X-rated horror fests.

But I had a good experience. Bit of local anaesthetic. Then a pushing, probing, inserting-a-needle-between-your-vertebrae sort of a feeling (I distinctly remember being warned to 'stay absolutely still' and wondering what would happen if one of my occasional body spasms kicked in), then a 'How does that feel?' (which I assumed meant 'Can you still feel your legs?'), and that was about it.

Despite the ease of the test, it almost goes without saying that the results – basically looking for signs of possible MS – were normal. A complete waste of time, except to confirm the increasing reasonableness of the hypothesis that I had PLS. The same could be said of my repeated MRI scan – once again, normal. This, rather regrettably, left me reluctantly volunteering to test my pain threshold, or as my test schedule euphemistically listed it, 'Repeat EMG'.

At least this time I was prepared for what was coming.

And as the young doctor ushering me into her laboratory immediately demonstrated, she was happy to talk. Without thinking about it, as usual I began chatting with her as if we were talking peer to peer. What had got her into neurology? How was she hoping to develop her career? How sensitive was the EMG equipment? Any professional topic to steer away from that overwhelming feeling of Me As Patient.

Thus it was that I rapidly built a rapport with the doctor responsible for my comfort over the next forty minutes. As she gave a running commentary of my test results, we discussed the science involved. Which was a lot better than simply waiting in stony silence for the next electric shock to come.

'Aah, bit of denervation here.' She was staring at the needle rammed into my calf. Basically, that meant that the single nerve running all the way from the junction box in my spine to the muscle in my lower leg had a problem with its insulation. It was breaking down a bit.

'Fascinating!' I replied. Well, it was fascinating. Because the insulation was supposed to be fine. It patently had been months earlier when I first suffered an EMG. We carried on chatting about attenuation and myelin sheathing.

'Hmm, looks like another site.'

'Really? Well, that's very interesting then. It's definitely indicative of a change in the last months.' I wasn't faking a calm exterior, I was genuinely fascinated in an almost objective way. That said, another part of me was having an inner dialogue along the lines of: 'Oh! Didn't expect that, did we? PLS is, by definition, now ruled out. Which leaves the ultra-nasty ALS. Probably. Not yet definitive though . . .'

I carried on chatting while she carried on shocking and nothing much progressed for about ten minutes. No more denervation. Until we reached the lump of flesh between my right thumb and forefinger. She stabbed it with her steely needle and wiggled it around a bit as she listened to the static emanating from a loudspeaker somewhere. I knew that the sooner the noise was good and loud the better – because then she'd stop wiggling that infernal needle.

If only to take my mind off her repositioning of the electrode, I demonstrated my limited anatomical knowledge with the words: 'Ah, the first dorsal interosseous! That's a common one to go ...' Because it is. It's a classic sign in more advanced ALS that you can see a long gap between the thumb and forefinger where the FDIO muscle has wasted away.

'Yes,' she said, 'and sure enough ... yes, definite denervation.' So, that was that then.

Without having time to process anything but the obvious conclusion, I immediately replied: 'Aah. So, according to El Escorial, incipient ALS.' El Escorial is shorthand for the internationally agreed criteria for diagnosing ALS. For example, if in addition to signs of upper motor neurone deterioration you also show lower motor neurone deterioration (such as denervation revealed by EMG) in three or more different sites, then it's concluded you have ALS.

Without hesitation, she simply said: 'Yes, exactly.' Just as if I were a medical student checking a diagnosis. Which was precisely what I wanted.

One second. Two seconds.

'Oh my God! I shouldn't have said that! I'm *terribly* sorry! Are you all right?'

I spent the next minute profusely reassuring my guilt-stricken doctor that I was perfectly OK. And lying to her that I'd expected the result and had merely been seeking confirmation. After all, she'd done exactly what I'd wanted her to do. And for that I was genuinely grateful to her.

The diagnosis of ALS not PLS was certainly a big surprise. But I'd already outlined in my mind what I believed was needed to survive the disease (namely feeding tube and ventilator pump) as well as what would be needed to lead a productive life with the disease (namely oodles of supercool hi-tech).

As I lay back on the examination table, and my unintended diagnostician completed the EMG, I remember thinking two things. One, I'd better start fleshing out the details of what hi-tech I was going to need. Two, how could I best explain the diagnosis to Francis to avoid upsetting him or unduly delaying our pre-arranged visit to the Ancient Egyptian Petrie Collection located a few blocks from the hospital?

Fortunately, thanks to our many chats and his own online detective work – and, to be fair, mainly because he's amazing – Francis was as matter-of-fact as I was. As we walked to the Petrie Museum, we agreed that my diagnosis *was* a Big Deal. But not *as* Big a Deal as everyone thought it was.

With that sorted, we spent the next two hours luxuriating amongst the tens of thousands of artefacts secreted in a couple of upstairs galleries within an almost hidden part of the University College campus. We enjoyed it so much we even bought the exorbitantly expensive catalogue. All in all, it was a *highly* educational day. And, in its own bizarre silver-lining way, a perversely lucky day too. I remember thinking as we arrived back at our nearby hotel: mine must have been one of the least traumatic diagnoses of ALS in history.

That sense of hubris lasted exactly nine and a half hours . . .

Life Choices

My stiff, slow, zombie-like walk back to the fencing salon must have made me look much like Connor. There was no blood on my white britches but I was in immense pain nevertheless – I felt traumatized. Even as I walked, I found myself wondering if I'd have preferred to be brutally caned like him. His humiliation was at least now over; mine seemed never-ending. The more I analysed my punishment, the more the framework holding my privileged life together continued to collapse. Nothing held true any more. Key elements of my future no longer existed.

The changing room was empty. Only my clothes, thrown down in another age, taunted me with my earlier carefree existence. I began changing, my mind elsewhere. My housemaster, sidling out of the headmaster's study straight after me, had headed in the direction of the out-of-bounds common room, turned, recommended: 'A little less *flamboyant*, Scott!' and disappeared.

'Ah, I thought you might be back.' It was the maître d'armes. I couldn't read his face. Irritated? Embarrassed? Bit of both? 'I'm sorry about the head of fencing ruling. I know it must be a huge disappointment and all that. But you have to see it from the school's point of view, the signal it would send to the other boys. And the school's reputation. Bit old-fashioned, I know. As far as I'm concerned, what men get up to on Hampstead Heath is up to them.'

I was so pleasantly surprised that my vigorous response was instantaneous.

'Yes, Maître! Absolutely! Of course!'

'But, as you're learning, I'm afraid not everyone thinks like me.'

He stood for a moment, wondering if he had anything else to say, decided he didn't, reached into his pocket and held out a key.

'Lock up when you go, please.'

This I took to be a parting gift; only the head of fencing was usually entrusted with locking up.

'Thank you, Maître.'

I finished dressing, threw my fencing bag over my shoulder, locked the changing room and walked around the playing fields to the entrance to the Slips. This was a narrow public right of way that cut through the school grounds, with a long high fence either side of it. It was my shortcut home. I looked at my watch. It was around a quarter past five.

Three minutes later my journey was over and I opened the glazed double doors to the entrance hall of the 1930s apartment block that had been home for all my life. I took the ancient lift to the top floor and let myself into my parents' flat.

A quick peck on my mother's cheek, then I used the excuse of three hours' homework to escape to my bedroom. My mind was still a whirlwind. I forced myself to concentrate on my physics prep and managed to lose myself in my work, only to be snapped back into my shattered reality. My father waved good evening as he headed to the main bedroom to change out of his city clothes. He worked at a small venture capital firm and commuted in a Daimler. His return home was a signal that dinner was about to be served. Soon after, I heard him pass my door again, and so I followed him to the dining room where my mother was about to plate up our meal.

She had once trained to be a doctor but had given that up when she married my father. For as long as I could remember, she had marked her subsequent days as a housewife by means of regimented menus based on vegetables collected from Wimbledon Village and meat delivered from Harrods food hall. It was Wednesday, which meant New Zealand lamb chops, mashed potato, carrots and peas.

After dessert, but before they turned on the Roberts wireless for the evening, I had something to ask them. Maybe, I hoped, to tell them. Perhaps it was the shock of recent events that emboldened me or even forced me to test something which I'd previously never dared, but also never needed, to go anywhere near. Now it was different. The day so far had tilted my world and I needed to know where its new axis was.

'We had an interesting debate at school today. For and against. On the one side: "Homosexuality is a perfectly acceptable lifestyle." On the other side: "Homosexuality is an abomination against God and humanity." I wondered what you both thought. There were strong views on both sides.'

My mother immediately put on her worried face. When she felt I needed her, she was always a deeply loving and supportive parent. Then again, so was my father. Both of them frequently hugged and kissed me. It was how they were. They now both immediately clicked into protective mode.

'Were there really, dear? On *both* sides? I find that very disappointing, don't you, Da?' She was looking expectantly at my father.

'Well, yes, I suppose so.' She kept looking at him expectantly until he added: 'I mean, yes, definitely.'

'I hope none of the masters expressed views on both sides, dear!'

'Well, actually, they did a bit.'

'I am horrified. *Horrified!* Really, that is so unfair on some of the boys. It gives them totally the wrong idea. Maybe I should ring the headmaster tomorrow and complain.'

'Oh, no! It's really not worth it. It was only a debate.'

'Well, if you're absolutely sure. But I think it's absolutely criminal exposing boys to that sort of talk. It could really harm them.'

'Well, I suppose it could be a little distressing if you didn't have the sort of support at home that I do.'

'Why, dear? What's happened?'

She was always fiercely intelligent and quick on the uptake. Sometimes *too* quick on the uptake. And all of a sudden we were skirting right along the very edge.

'No, no, nothing! I was just meaning that otherwise some boys might find it very confusing.'

'Well, you are in safe hands!'

She relaxed. I relaxed. My father almost always seemed relatively relaxed. She reached out her hand and rested it on mine.

'Never, ever worry.' She was at her most reassuring. 'There is absolutely nothing to feel confused about. As Da will confirm, there is *no* question about it. Of course homosexuality is an abomination. Isn't it, darling?'

'Of course it is!'

'Rest assured,' my mother patted my hand, 'no one in all of the extended family would ever have anything to do with someone who was homosexual.' She took her hand away, pulled her arms to her chest and gave an exaggerated shiver. 'Oh, the shame! I mean for the parents. It must be mortifying for them, with everybody pitying them. Especially the mother. The latest research suggests it's a domineering mother and weak father that is the cause. How humiliating! Don't you think, darling?'

I pointed out that I must finish my prep, smiled, got up, went to my room, locked the door and cried. I cried at the unfairness and the cruelty. I cried at the loss of the familiar world I'd known at a school where I'd flourished, where I'd held my head high. I cried at the certainty that my own parents not only didn't fully know me but regarded people like me as an abomination.

That night, I hardly slept. My mind was processing. Reassessing. Discarding. Anguishing. I vividly remember being awake and watching the dawn a bit before 5 a.m. At the end of that long night, I had made up my mind.

I would follow my headmaster's advice.

Thirty-six hours later, I was emerging with Anthony from after-school choir practice.

'New look!' He made it sound like an American TV commercial. I struck a pose. 'New hair *and* new suit.'

I now sported a centre parting and was wearing a stylish navy-blue suit with large-peaked lapels and very wide flares.

'I'm going to grow it long. And how about the heels?'

I pulled on my trouser leg to reveal two-inch heels. They pushed my height to well over six foot. Anthony adopted the exaggerated persona of my housemaster.

'The shame! The shame! What *are* you doing, boy?'

'I'm doing *exactly* what the beak told me to do, Anthony. I'm starting a new life.'

'Well, it looks great.' He sounded totally genuine.

'Oh, this is just the beginning. I've got rather a lot planned.'

'Ooh! Can't wait! I want to know every minutest detail. You can tell all at fencing tomorrow.' He paused, and for once used his real voice. 'I really am so sorry about the head of fencing thing.'

This seemed as good a time as any to bring Anthony up to

date on the latest. We were still only halfway down the long main driveway of the school – enough time to give him the highlights before we split in different directions along the Ridgway.

'Actually, I'm not going to be at fencing tomorrow. I'm never going back.'

'*What?*' He stopped walking.

'For the last couple of months, I've been going some evenings to this amazing dojo – that's Japanese for "school" – run by a guy who is the highest-graded black belt outside of Japan. It's the toughest form of karate there is: Kyokushinkai. Full contact. You don't pull your punches. Or kicks.'

'And?'

'And I've persuaded the games master to let me do ten hours a week there instead of fencing. I explained there wasn't time to do both. I then convinced him by saying my only regret was that I would no longer be able to represent the school in fencing matches! Perhaps he needed reminding, but that clinched it. He was positively encouraging me by the time I'd finished.'

'Serves them right. So, where is this Palace of Pugilism?'

'Raynes Park. It only takes twenty minutes by bike. And there are *real* people there. I'm sure I'm the only public schoolboy. Most are older anyway. I love it.'

'*Very* Bruce Lee . . .' Martial arts films were all the rage, and *Enter the Dragon* starring sexy Bruce Lee was the absolute gold standard. Anthony started walking again. 'And what about dram soc? What next for your curtailed acting career?'

'All the world's a stage, dear boy!'

'And all the men and women merely players, dear boy! Yes, I know. But if not dram soc, what?'

'Computing! Computers are the future.'

'*Computers?*' He stopped walking for a second time.

'I'm joining the computer society.'

'Don't tell me: we don't actually have one but you're joining one in New York?'

'We *didn't* have one. But Mr Billings has got access to an IBM computer a few of us can learn on once a week.'

'Where on Earth is that?'

'Merton Polytechnic.'

'*Polytechnic!*' He reverted to housemaster mode. 'The shame! The shame! They don't learn Latin. They split their infinitives. They can't even *spell* Oxbridge . . .'

'I know, isn't it wonderful? I might even fall in love with a bit of rough!'

'More likely, a gang of them will rough you up, Peter!'

'Hence the karate . . .'

A broad smile slowly spread across Anthony's face as the logical connections between my sexuality, my newfound rebellion and my change of sport became clearer.

'The school doesn't know what it's started!'

'Seriously, I feel unchained, liberated. I'm going to completely reinvent myself.'

And as I said it, I knew it was true. Anthony's enthusiastic response gave me valuable and much-needed belief and confidence.

'Metamorphosis!'

'*Yes!* Metamorphosis. That's the word. The Establishment doesn't like who I was; just wait till they see what I'm going to turn into! They'll never have seen anyone like me before. They won't know what's hit 'em!'

'You're not going to take on the whole school?'

'No, of course not.' I paused for dramatic effect. 'I'm going to take on the whole world! You know, the maître once said to me: "Peter, always aim for the stars; that way at least you'll reach the moon." The more I think about that, the more I realize I have absolutely zero interest in exploring a grey rock. I'm engaging warp drive!'

This time, Anthony said nothing. He just stopped and stared quizzically.

'From now on, I refuse to tolerate what's unfair in the status quo; I'll change it. I refuse to be beaten into submission and forced to conform by having options taken away from me; I'll turn liabilities into assets and create *new* options. And I absolutely refuse to give in to bullies – however official they are. From now on, every time the Establishment tries to bully me, I'm going to push back, and back, and back. Until eventually, they give in.'

'And you get a negotiated peace!'

'No! Until I get unconditional surrender!'

Anthony was chortling. He clearly loved the idea. As he began walking again, he summed up his delight:

'It's going to be a battle royal!'

I stopped. He stopped.

'Battle?' I cried in mock incredulity. 'This is *war!*'

The Morning After

I jolted awake and squinted at the unfamiliar bedside clock: 03.05. I was instantly fully aware, slapped by the instant realization that only upon waking was I having a nightmare.

It was the sense of impotent dread I would forever associate with being sixteen, waiting outside the headmaster's study, listening with a white face, open mouth, sick stomach and hammering heart to the muffled sounds coming through the padded door – murmuring, silence, one stroke, two, three, murmuring, door open, 'Next!'

Without warning, what felt like a night terror attacked. Followed by another. Until they overwhelmed me.

You're going to die!

No, I'm not. I'll have a feeding tube and a mechanical ventilator.

Pathetic answer – those will not be enough. If they were, everyone with MND would do the same. But they die instead. Just like you will. Unlike you, they have the courage to accept the inevitable.

It's not inevitable. People just don't often choose to carry on until they're locked in.

Don't be stupid, you delusional, pompous fantasist! You're a disgrace to Science. Trust the statistics – you'll most probably be dead in two years. Like everyone else.

If Stephen Hawking can survive, so can I.

Aah, here it comes. The great Dr Scott-Morgan compares himself to the most famous cosmologist on the planet. How arrogant, how desperate is that?

He had a vent pump fitted in 1985.

He got MND when he was much younger than you. He deteriorated far more slowly than you. He is much richer than you and can afford the best 24-hour care. You are not, and never will be, in any possible way, comparable. You will die a common, average, unexceptional death. And no one will notice.

Even if I do die early, I should be able to last at least five years. Ten per cent of patients last five years.

Yes, but you don't know if they have less aggressive forms of MND than you. You'll probably deteriorate faster. Even if you do last five years, you'll be completely paralysed apart from your eyes – powerless, trapped for as long as you can survive in the ultimate straitjacket of your own living corpse.

Other people cope.

You won't. All your senses will be unaffected. You'll feel every itch but never be able to scratch.

Other people cope.

You won't. You'll get claustrophobic. Remember when you got stuck crawling through that tight cave tunnel as a fearless undergrad? You've always prided yourself on how you calmed yourself and thought your way out of it, but you won't be able to think your way out of this one. Your clever brain and all your expensive education won't make the slightest difference, will it?

No.

You won't even be able to cope with not talking. And everyone who's ever had to put up with you going on and on and on interminably will be delighted you've finally shut up. And you won't be able to stand it.

I'm not sure . . .

You're about to destroy the lives of everyone around you. You're not the only one who won't be able to stand it all, you know. You only got your first symptoms a year ago, yet you're already a cripple. People look away from you as you shuffle towards them. You're an embarrassment. An embarrassment to Francis.

I know.

He didn't sign up for this.

I know.

Yet despite that, you have such an overinflated ego that you insist on single-mindedly pursuing your own self-centred agenda of survival at any cost — oblivious to the collateral damage to those around you, to the person who loves you most.

Yes.

He deserves better.

Yes.

If you love him a fraction as much as you claim, you won't prolong his suffering. You'll save him from having to watch you slowly decay into uselessness, continuously saying one painful goodbye to a shared activity after another until eventually you go to a place where even he cannot follow. If you really love him, you'll protect him.

Yes.

Because otherwise, he'll learn to resent you. And then he'll leave you. And he'll put you in a nursing home full of old people that smells of piss. And he'll let you die alone.

What?!? This is absurd. It's the middle of the night, and it's the first time my subconscious has had a chance to process the full enormity of my diagnosis. Take a deep breath. Calm down. Think your way through this.

Gradually, I became aware of the reassuring snuffle of Francis softly snoring beside me. Always beside me. Whatever the battle. Francis and Peter Against the World. I could no longer feel my heart pounding, my breathing had returned to normal and there were no terrors lurking in the shadows.

In their place was something far more pitiable.

As my subconscious and conscious persisted in their internal monologue for a while longer, our interaction seamlessly transformed into two completely different characters. My co-actor now appeared to be little more than a boy, years before he would be forced to wait outside the headmaster's study. He was hiding in a corner, naked on the floor, hugging his knees and shivering, just as I was from the still-cold sweat.

I'm scared.

I know you are. It's all right to be scared. It's perfectly reasonable. But in the morning, the sun will come up, and Francis will be awake, and we'll have a nice breakfast, and we'll all feel a lot better. And then Francis and I will take on the world, and win. For now, we've got to be strong.

But MND is stronger.

No, it's not! MND is just another bully, the ultimate Establishment bully. And the Scott-Morgans don't give in to bullies. You know that.

But there's no treatment.

There are lots and lots and lots of possible treatments. It's just that they're hi-tech treatments, not medical treatments, so their potential is being ignored. MND has intimidated people unopposed for so long that its unquestioned reign of terror has become institutionalized. Everyone's just waiting for a magical cure to save them. But I believe in the possibility of something else, something radically new, something utterly amazing. With technology no one has seen before. Science fiction made real. I know it's possible.

Will it be fun?

What?

I was completely unprepared for my brain to take such a violent tack as it veered course. For ten seconds, maybe twenty seconds, I luxuriated in the exquisiteness of experiencing polar-opposite emotions at the same time: still fear, anger, despair but also – just as powerfully now – excitement, joy, hope.

Then, the new emotions, the positive emotions, began to

dominate. I felt a warmth, a sense of *power* spreading out from my inner core. The last psychic echo of the terrors dwindled into irrelevance, vanquished.

With my cheeks still prickling from drying tears, I found myself smiling. I felt exuberant.

It's going to be a bit 'best of times, worst of times' but it's going to be awesome! And we'll have to track down some of the most ultra-cool hi-tech in this corner of the galaxy.

It's an adventure! We love adventures!

We've got two years before statistically I should be dead. That means we've got two years to rewrite the future. And change the world.

There'll be battles all the way, and then the ultimate life-or-death showdown. Either we'll win, in which case *everything* will change, or we'll very conspicuously fail. Which isn't going to happen. There can be no middle ground.

MND expects me to die.

I refuse.

I also refuse merely to 'stay alive' in a form of living death.

Also – a complete revelation to me as the thought distilled and suddenly became recognizable – I refuse too to leave everyone else behind, traumatized by their two-year death sentence, scared to die, terrified to live. We'll gather an army. We'll build a movement. This is rebellion!

And this isn't just about me and about us. This is about using cutting-edge technology to solve other forms of extreme disability, caused by disease, or accident or old age. This is about everyone who's ever felt themselves to be a freethinking intelligence trapped in an inadequate physical body. This is about every teenager – and grown-up – who's ever wanted to be more, better, different . . .

This is about *changing what it means to be human*.

I'm not going to waste another second working out how I can stay alive. I'm now not the least bit interested in ways to survive. Tonight, Francis and I are going to crack open our *very best* bottle of champagne and we'll celebrate that I am *not* going to solve the problem of how to survive.

I'm going to solve how we can all truly *thrive* . . .

Bicentennial

I woke again when my head knocked against the window next to my seat as the Greyhound bus sped over another pothole on Highway 1. The cheap air pillow I'd bought thousands of miles earlier had deflated again. I found myself smiling; I'd been dreaming of Brad.

It was early dawn and I could see glimpses of water, so I decided I might as well stay awake and watch the sunrise. I carefully stretched, trying not to wake Anthony, whose head was resting on my shoulder. Each night we alternated who got the window seat.

We'd been travelling like this for almost two months. It was the summer of '76 – Bicentennial Year, when all the fire hydrants in the USA were painted red, white and blue. We'd started in New York, were in Philadelphia for the huge 200th Fourth of July celebration with President Ford, then travelled by bus in a huge circle across the continent to the West Coast, down and back again, and had recently experienced every theme park in Florida and, far more important to me, Cape Canaveral (still Cape Kennedy as far as I was concerned, forever associated in my mind with the Apollo moon shots). We were now heading to the southernmost point of mainland USA: tropical Key West.

Slim young Brad had sauntered on to our bus at St Louis wearing tight jeans over cowboy boots, a tight cap-sleeved T-shirt and a genuine cowboy hat. He'd hoisted his holdall on to the luggage rack and sat across the aisle from Anthony and me. Fortunately, it had been Anthony's turn for the window seat.

After a hundred miles or so, with the sun now set and the view obscured, Brad and I had started chatting. When Anthony said he'd try to get a bit of shut-eye, I'd scooted across the aisle to give him a bit of peace. Brad was nineteen, a year older than me. Unlike me, he'd performed in rodeo and could ride a real bucking bronco. He was an incredibly cute All-American Boy. And half an hour after sitting next to him, I'd realized his hand was beginning to touch my leg.

With the bus in convenient darkness and Anthony in convenient slumber, Brad and I soon started kissing. It was the first time in my life I had ever kissed another guy – apart from members of my extended family (and that hadn't involved tongues). A true cowboy, Brad had kept his hat on at all times.

He'd left the bus in the middle of the night at Kansas City. Standing like a statue outside my window, he'd waited until the bus pulled away again. Then, with a half-grin and slight nod, he touched the rim of his cowboy hat in silent farewell and stood watching until the bus turned a corner and he disappeared from sight. Only then did it strike me that, in all the excitement of my first proper kiss, I'd never even thought to touch his leg back. Let alone anywhere else below his slim waist.

It smelled like we were passing a sewage farm, but with ocean now visible on both sides of the highway, this seemed unlikely. I elbowed Anthony awake.

'You farted!'

He sleepily reached for his glasses.

'Where are we?'

'I think we're approaching the Seven Mile Bridge. It's one of the longest in the world, so it's worth seeing. Anyway, look at the dawn!'

In the last few minutes, the sky had turned coral-orange

and the previously grey sea was suddenly deep turquoise. I'd never seen anything like it.

'I've never seen anything like it,' echoed Anthony. 'And I formally disavow any knowledge of your slanderous accusation of malodorous action.' He was taking a law degree, after all. We shared the cold remnants of coffee from the thermos last filled in Miami. This appeared to kick-start his brain: 'Have you been back to King's since you left? Looked up your old friend the headmaster, perchance?'

'No, I've never again darkened the doors of those hallowed halls.' I watched the coraled sea speeding past, then remembered: 'I turned down two invitations from masters kindly offering me an explicit form of extra-curricular education – you'll never guess one of them! And I *have* had intimate boozy dinners with three others. I'm becoming rather good at my "lovely dessert but I won't have any sex with the brandy, thank you" speech . . .'

After Anthony had extracted every gossipy morsel, he ruminated for a long while, then stated: 'I still don't get your machine intelligence problem!'

I couldn't help thinking that if it had been the convoluted plotline of a Wagnerian opera he'd have got it in an instant, but I persevered anyway.

'Look, it's like this. As I've explained, *many* times' – at this, he had the good manners to look suitably sheepish – 'computers will get smarter and smarter until eventually you won't be able to tell whether the answers to your questions are being typed by a human or a machine.'

'The Turing test!' Anthony enthusiastically exploded, if only to demonstrate that he really had been listening to me through the last several states.

'Exactly – the Turing test. Now, everyone assumes that soon afterwards the intelligent machine (or maybe several intelligent machines) will be able to construct an even more intelligent machine. And so on.'

'Yes, yes, I get all that. The classic route is that machines become super-intelligent and eventually take over. But *you*, in your infinite wisdom, want them instead to be –'

'Intelligence amplifiers!' I broke in.

'Right. You think it's better that humans *merge* their brains with machine brains as the next step in our evolution, extending our IQ *and* lifespan indefinitely, travel to the stars, populate the universe, all that science-fiction stuff. OK, I get it. So, what's the problem? Apart from the fact, of course, that the best computers are currently about as smart as a slug.'

He'd already demonstrated vastly more knowledge of computing than I had of opera (even after his extensive tutelage), so I let him down gently.

'I'm not sure they're even as bright as a slug, but leaving that to one side . . . My problem is that the few scientists who support the intelligence amplifier route have it all wrong. They go around suggesting that one day we'll be able to scan our carbon-based biological brains, transfer the data to silicon-based computer brains and from then on break free of our human lifespan.'

'And what's wrong with that?'

'It won't work!'

'No, but none of it works now. Surely it's only a matter of time?'

'I mean, it *can't* work. It's theoretically impossible.'

'How can you *know* that?'

'Because it's a *copy*. The "immortal" silicon brain they glibly talk about is a *copy* of the original. *It* will have all the memories of the original, and *it* will potentially live forever, but if you're unlucky enough to be the original, you die as normal. Worse, if the scanning process destroys your biological brain neurone by neurone, you'll die as a result of the transfer itself! *That's* what I call the Uploading Problem.'

'Aah! Is that why you were wittering on yesterday about Captain Kirk being killed every time they beam him from the *Enterprise*?'

'*Yes!* It's exactly the same thing. The transporter scans the body as it dematerializes it, transmits the data through space, then reconstitutes the body on the planet below.'

'I *have* watched *Star Trek* . . .'

'But even when I was a kid, it was obvious to me that they were reconstituting *copies* of Kirk each time. If they didn't dematerialize the original, they'd end up with *two* Kirks – one on the *Enterprise* and one on the planet surface. Sure enough, in one episode they *did*. So even the writers knew they were killing Kirk and his crew in every episode.'

'It must have been very expensive getting all those actors –'

'Oh, look! It's a pelican!'

We were both entranced by Key West. Even the bus station was a huge upgrade on what we'd become used to. We'd long ago realized that one of the many hidden advantages of exploring the USA by Greyhound was that every city terminal tended to be in the cheapest (as in, poorest) part of town. The monuments and tourist attractions, in contrast, tended to be in the most expensive (as in, affluent) part of town. To walk from one to the other involved traversing the full economic spectrum the city had to offer.

'Where are the tramps?' Anthony asked in a bemused voice. He was referencing Washington, where we'd spent much of the day visiting the White House and the Lincoln Memorial but ended up in what – at least to two boys from Wimbledon – looked like slums.

'I'd just bitten into my burger!' I reminded him.

'You didn't have to spit it out.'

'He was peeing against the café window inches from my face!'

A few hours later, having moved from the Key West drop-off to Old Town, I was falling in love. 'You know, I could really live in a place like this.' We'd been lent a brightly coloured tandem by some locals who had insisted we properly explore their lovely island. We'd just reached Key West Harbour and were now toying with the idea of having a drink at Sloppy Joe's, where Hemingway used to hang out.

'I'd love to live in the States, certainly.'

'Wouldn't you have to learn US law if you practised here?'

'Oh, I'm only doing law to reassure my mum that I've got a second string to my bow. It's always been opera for me. *That's* my passion.'

'Of course it is. And of course you're right. You love opera. I'm sure you'd be good as a barrister but you'd be great as an impresario, and it's inevitable you'll choose opera over another "more sensible" career. My latest Law of Logic and Love predicts it.'

'It's nicely alliterative, I'll give it that. What does it say?'

'It's an observation of how I think the universe works. I believe it's an inviolable Unwritten Rule when it comes to how each of us makes life-changing decisions: Logic gets us so far, but whenever it comes to a showdown, True Love wins over Logic every time . . .'

Seven weeks after returning to a long, hot summer in the UK, Ma and Da drove the Daimler and me, and lots of bags and boxes, into London's sumptuous South Kensington, the (very convenient) location of the Imperial College of Science and Technology.

In an act of supreme rebellion (and, in the words of my former housemaster, 'supreme stupidity'), I had turned down Oxford or Cambridge in favour of Imperial – which to my mind was a perfectly logical choice as it was the only university in the country to offer a full-time undergraduate course in computing science. I was now convinced that computers were the future; my housemaster, and, according to him, 'anyone with even a modicum of intellect', was equally convinced that my flagrant disregard of the Unwritten Rules of the Establishment by willingly selecting anywhere but Oxbridge was 'a betrayal of the school, your parents, and above all yourself'.

'It really is a magnificent location, darling,' Ma passed verdict. 'Right in the West End! Across the park you can walk to Oxford Street. Past the Serpentine, you're in Kensington Gardens. The other end you're in the theatre district. And of course you're *feet* away from Knightsbridge, so if you ever run low on food, you can always pop into Harrods food hall.'

We found my good-sized single room in one of the student accommodation blocks and began to transfer my belongings. That done, we brewed a cup of tea using my new electric kettle and old teapot, washed the cups in the small basin next to the bed and agreed that now was a good time for them to set off back to Wimbledon. I walked them to the car, kissed them goodbye, waved as they drove away, introduced myself to my bedders (two motherly Cockney charladies who would make my bed and clean my room for me every day) and returned to my room.

I stood by the desk, looking out through the large window on to the gardens below. A huge smile was stretching my face to the limit. I had escaped . . . I was no longer going to be surrounded by the cloying conventions, bigotry and 'keeping up appearances' mentality of upper-middle-class suburbia. I had escaped from the privilege and brutality of a school I had once loved and an Establishment system I had grown to hate. And I had escaped from parents and an extended family that claimed to hate people like me but, because they didn't know one of the most important things about me, also claimed to love me. My childhood and upbringing gave me opportunities for which I could only be intensely grateful. I would never deny that. But at what cost? What value does opportunity have when it comes with the condition of not being who you are, who you were born to be?

I suddenly let out a very loud whoop of elation. Then I turned away from the window and, flushed with hard-won euphoria, headed for the campus.

I was free.

Bollocks to That!

'If there's one thing I've learned, it's that if you are going to be diagnosed with a terminal illness, MND is the one to go for . . .'

Helen, an old friend of almost thirty years, from a time when I'd worked in Berkeley Square and she'd been my MD's secretary, laughed, then wondered whether she should have. Then she laughed again, took another sip of her latte and raised her eyebrows for me to elucidate.

'Seriously!' I enthused. 'There's no chronic pain, there's no nausea and – unlike people who have a sudden accident or have a brain tumour or something – we have *plenty* of time to order our affairs. And . . .' I paused for effect, '. . . critically, it turns out that the terminal bit is . . .' I still struggled to decide the word people would respond to best for the next bit, '. . . negotiable. It's completely up for grabs.'

We were waiting in a so-called restaurant in an NHS London teaching hospital prior to my consultation with the head of the MND clinic. It was set to be the first time my self-diagnosis was officially confirmed face to face – although unofficially he'd confirmed it by email when I contacted him to alert him to the test results he'd be receiving and their obvious implications.

'But isn't MND usually claimed to be The World's Cruellest Disease?' One of the wonderful things about Helen was that she wasn't afraid to go where other angels feared to tread.

'Yes, but at the moment I'm not sure I understand why. I'd

far rather have this than a fatal brain tumour. Or chronic pain. Or feeling sick all the time. Or losing my mind. Or a whole range of horrible endings.'

'I agree, but in that case, why do they say it?'

'I don't know! Yes, it's awful, and I have no doubt how much suffering it can bring to those who have it – as well as those around them. But I just feel it *could* be much worse. And hopefully, this guy I'm about to see will be key to helping me stay as far ahead of the curve as possible as my body shuts down.'

Looking back, I suspect Helen thought I was in denial. 'Surely *everybody* wants that. I assume he'll give you a leaflet or something. The NHS is really very good on this sort of thing.'

'I refuse to become involved in any of this!'

As friendly and charming as the conversation had appeared to be for the first five minutes of our consultation, my doctor was now getting rather hot under the collar despite the perfectly adequate air-conditioning.

'Really? It's just that I want to be steadfastly proactive in my overall clinical care. As I just said, I intend to always remain well ahead of the curve . . .'

'I repeat, I refuse to become involved in that,' he said, surprising me with his anger. 'MND follows no rules. You *cannot* be proactive. You cannot be anything other than *reactive*!'

I wanted to call him a total arse but instead politely answered that I found that a very interesting viewpoint, but was there no way that I could at least tap into his unparalleled experience to anticipate the most likely range of scenarios of how my *current* condition might evolve, so that I could best prepare?

'Absolutely not!' Followed by the insightful clarification:

'After all, if we did that for you, we'd have to do it for *everyone* with MND.'

Well, that would be an interesting experiment – personalized clinical care. With my gritted teeth fixed in a rictus smile, I persevered with a strategy of diplomacy. It was a total waste of time and breath. My reluctant conclusion twenty-five minutes later was that one of the most experienced MND clinicians in the country might indeed be a top diagnostician but he didn't have a scientific bone in his body. The status quo. simply existed. There was no way of breaking it. No point in even trying.

Not only was he actively against my whole idea of being proactive but his verdict on my worsening clonus (both my legs could now vibrate uncontrollably sometimes) was: 'You're massively under-medicated!' He'd promptly written out a prescription for the largest dose allowed outside of hospital supervision of a muscle relaxant called baclofen. That was very kind of him. But I questioned the logic of dramatically relaxing the muscles of someone already struggling to get their muscles to work properly at all.

With the aid of my trusty cane, I hobbled from the consultation room towards the main exit. So, this was it. I was supposed to wait and see. I was supposed to succumb to the inevitable and conform with the established protocol. I was supposed to accept the medical dogma that there was no effective treatment for MND and that it was fatal. As I reached the glass doors and they hissed open and cold air punched my face, I found myself speaking out loud my profound thoughts on the matter:

'Bollocks to that!'

Changing the world of MND – let alone changing what it means to be human – was not going to be straightforward.

'This is going to be a bloody pain,' I commented to Francis after relaying the highlights of my earlier conversation.

'Of course it is! The medical profession is as driven by Unwritten Rules as any other.'

'Actually, it's even more driven by Unwritten Rules than most corporations. I remember an analysis we did of the NHS and a range of hospitals in the USA and –'

'That's my point! You're the world expert on Unwritten Rules and how they drive the future. You *know* how these places work. And if you don't know, you know how to work it out. So, if you're so bloody clever, use your expertise to work out how to totally rewrite the future of shit like MND. It's what you always told your clients you could do for them. So prove it. Do what you do, but this time, do it for you. And everyone else who'll benefit.'

Francis was right, insofar that during my professional career I'd worked out how to decode and potentially transform the hidden inner workings of everything from organizations to global systems, and had authored a few books on the subject. After such a blatant appeal to my professional pride, I felt honour-bound to step up to the challenge. I politely fired my consultant in London and transferred myself back to the tender ministrations of the NHS in Devon. I was beginning to realize that there were potential advantages to being away from the sprawling metropolitan teaching hospitals where professionals could be so concerned about the risk of reputational damage that they sometimes sat on their laurels so often that the laurels eventually wilted. I had a hunch that, counterintuitively, the NHS in the West Country might be a little more inclined to experiment.

With this in mind, Francis and I ushered into our living room a lovely lady with the snappy title of 'MND Lead Clinical Nurse / Care Network Coordinator, Southwest Peninsula'.

Having already used up five minutes of our meeting reading her business card, I thought I'd cut to the chase.

'Tracy, it's really wonderful to meet you. Let me save a little time by giving a bit of context.'

She leaned forward on the sofa, picked up another biscuit, leaned back again and smiled expectantly.

'As you know, given my diagnosis date, statistically I will be dead within twenty-two months.' I'd caught Tracy just as she'd bitten into a chocolate digestive, but even amongst a shower of crumbs she managed to look suitably sombre. 'Now, I know you don't know me, but take a long hard look. In your intuitive professional opinion, do I look like the sort of person who's going to statistically curl up and die on cue?'

Tracy asserted her denial with a vehement shake of her head and an accompanying 'Mmm *mmm*!'

'Now, you know and I know that waiting for The Cure is a bit delusional.'

She gave a non-committal raise of her eyebrows, so I pushed home the point.

'Around the world, every MND charity, every instance of the Ice Bucket Challenge, is all about encouraging the public to cough up donations to fund medical research into finding a cure. Fifty years of research later, only *one* drug from twenty years ago, riluzole, *may* add a couple of months to our lives. Worse, however the charities and researchers spin it, there is *nothing* on the research horizon that can possibly be ready in time to help any of us that have already been diagnosed.'

'You know of the various trials that are going on?'

'Yes, of course, but at best some of them will slow the rate of deterioration, maybe. And nothing would be ready for, what, five years? Ninety per cent of the people diagnosed when I was will be dead by then. And a *cure* that would actually reverse the ravages of MND and rebuild my muscles and my motor neurones and the motor cortex of my brain may well be *decades* away.'

Tracy pursed her lips and slowly nodded.

'Obviously, therefore, I'm going to have to take a different route. To do it, I'll need your help in due course to get a feeding tube inserted into my stomach and a ventilation pump connected to my airway to keep me alive indefinitely. Both my parents were very healthy and lived to their nineties; I've never had a sick day in my entire career. Therefore, I think we should plan my clinical care on the assumption that I'll potentially be locked in for decades.'

For some reason, Tracy stopped mid-bite and for a few seconds was left sucking the chocolate from the biscuit before recommencing her mastication.

'Now, my doctorate is in robotics so, as you're probably thinking, with that background, what a fantastic opportunity to do some serious research!'

I peripherally registered that Tracy's face had frozen and hoped it wasn't through boredom. Just in case, I upped my enthusiasm.

'I plan to throw a huge amount of cutting-edge technology at this to see how we can make being locked in really exciting. To me, this is literally the experiment of my life!'

I smiled. Not a flicker back.

'In summary, I really look forward to a very long-term and fruitful relationship.'

I assumed everybody Tracy met for the first time started off with something like that. Interestingly, though, she wasn't especially forthcoming with her response. Indeed, based on observation of Tracy's face throughout the last thirty seconds of my introduction, I suspected she was very good at poker.

And then she came up with the immortal words: 'Well . . . that's *very* helpful.'

The conversation continued swimmingly for, as I remember it, about an hour until, with inevitability, the topic under discussion migrated towards bodily functions. It was around this time that Tracy first used a phraseology that I was subsequently to learn is almost obligatory for a clinical care professional to insert into any introductory meeting with someone recently diagnosed with MND:

'The good news is that throughout the course of the disease you are likely to remain fully continent.'

She said this as if it were a *good* thing. In reality, it's actually the set-up for the ultimate good news/bad news gag:

'The good news, Dr Scott-Morgan, is that you will always be able to go to the bathroom. The bad news is that soon you will never again be able to *reach* the bathroom.'

I'd actually realized this weeks earlier, so I'd come to the conversation with some ideas of my own. However, Tracy seemed very comfortable as things moved on to what apparently felt to her like home territory. So I thought I'd let the conversation run. Anyway, given that everybody with MND surely had solved this self-same issue, I was very interested to know what the standard solution was.

'Great! So, how does clinical care deal with the fact that when you're paralysed you can't get to the loo?'

'Carers,' she explained.

'OK, well, what do they do then?'

'Ah,' she answered as if it were the most natural thing in the world, 'they'll toilet you.'

Now, I don't know if you've ever reflected on what a resolutely *passive* verb To Be Toileted actually is. For some childishly irrational reason, I was not immediately enamoured of this scenario. Nevertheless, I persevered and continued with what I thought was the very obvious supplementary question:

'OK, so when I inevitably get pneumonia, and we treat it with a strong course of antibiotics and I get diarrhoea, what will the carers do then?'

Tracy was practically beaming now: she knew the answer.

'That's what incontinence pads are for!'

It was at this stage that I decided that maybe I would introduce some of the ideas I'd prepared earlier.

'Can I suggest that perhaps we do something very slightly different?'

Being Immortal

'I was shitting myself! Seriously, it was the most claustrophobic experience of my life. There I am, in this super-narrow cave tunnel, just big enough to crawl along with my elbows tucked in and my helmet scraping the ceiling, and *this* arsehole' – I tilted my beer glass towards little Nick who, when he grinned back, looked remarkably like Connor – 'calls out: "We've got to go back! It's too tight!" And I dutifully tried to oblige, and that's when I found that I *couldn't* go back, I could only crawl forward!'

I was holding court in the Students' Union bar, buoyed by the joy and exuberance of someone who has survived something immensely dangerous and lived to tell the tale. More than that, I was nineteen and having the time of my life. My now shoulder-length hair had been lightened and my skin tanned by being outdoors for much of the preceding summer and early autumn, I was wearing a white cap-sleeved T-shirt and under the table, if any of my four friends around it had cared to look, were skintight Levi jeans worn over cowboy boots. I referred to it as my James Dean look, but secretly it was my homage to my Greyhound cowboy, Brad.

'How come Nick wasn't stuck?' This came from John, who was already a pint ahead of the rest of us. He and I had spent part of the summer at Lasham Airfield getting our gliding licences.

'Because he's the size of a fucking mole, you wanker.' Buster, a year ahead of me, appeared to be celebrating the approach of his electrical engineering finals by squeezing expletives into every sentence he uttered. When he and I had, after a day's training at Thruxton, climbed out on to the wing of a light aircraft, let go and counted six seconds in a loud voice to see if we needed to deploy the emergency parachute hanging below our waists, Buster alone had screamed: 'Fucking one, fucking two . . .' until, no doubt to the relief of the gathered spectators below, his parachute finally opened on 'fucking five'.

'Fuck you!' retorted Nick, grinning his cute grin.

'You have to remember,' I resumed, 'I was freezing cold from swimming through the sump.'

'What the fuck's that?' Buster, of course.

'A flooded tunnel where you have to take a deep breath, swim through the flooded bit and come out the other side,' explained Nick, an experienced potholer. 'The water's always bloody cold, whatever the time of year.'

'So how did you get out?' John appeared to have a genuine though possibly alcohol-induced interest in the answer.

'Well, fortunately,' I continued, 'at that very moment, young Nick let off a substantial stream of methane.' Everybody broke down into uncontrollable schoolboy laughter.

'I was concerned that you were stuck,' justified Nick eventually.

'*I* was concerned that you had become a flammable explosive device! A single spark could have brought the whole tunnel down.'

'So *how did you get out?!?*' John persisted.

'Well, with all the incentive of clear and present danger, I just held my breath and got the bloody hell out of there!'

We all laughed and took a sip of beer, despite my last sentence having been a complete lie. Bravado hid how I had really felt. In reality, when I realized that I was stuck, in near-total darkness, with hundreds of feet of rock above me, with me blocking Nick from escaping, with no one behind me to pull me back by my legs, I froze. For the first time in my life, I felt genuinely, utterly terrified, crushed by claustrophobia; for the first time in my life, I felt absolute panic rising. It felt like my head was being inflated with boiling water. I was losing control.

My rational brain had struggled to assert itself. This is serious. If I panic, it could get dangerous. A few years earlier, I'd survived being almost electrocuted to death. Just. I'd been alone at school, after hours, both hands connected by some faulty wiring to the London electricity main, unable to move, not so much scared of imminent death as surprised by it, an alien black shroud of tunnel-vision irising shut, centred on the prosaic light-fitting about to kill me. I put my last efforts into setting the light slowly rocking, each time slightly more, trying to unbalance it, until I blacked out as it started to topple.

I'd thought my way out of that, hadn't I? This was the same. After a moment, I began to settle. Treat it like any other problem, I told myself. Be calm. Blot everything else out. Push the panic back down again. I had to *think* my way out of the crisis. I remember beginning to concentrate. Unfortunately, perhaps because he was attempting to wriggle backwards, perhaps because he felt himself to be literally in a tight spot, it was at the peak of my concentration that Nick *very* noisily let rip. That part of the story was painfully true. I won't say that it reverberated through the cave complex but, in the tomb-like silence of our stony coffin, it sounded *very* loud.

'Sorry!'

I resumed my concentration. I experimented with wriggling. It failed. I tried pushing back with my elbows. That failed too. Exasperated, I finally resorted to a complex triple

85

movement involving pushing with my elbows, pulling with the toes of my boots and arching my body. I moved slightly back. It had worked! A sense of relief, optimism. From then on, it was merely a question of repeating the exercise for what felt like a few hundred times, before claustrophobia finally gave way to elated relief.

'Well,' Nick had said as he emerged soon after me, 'that was easier than I'd expected.'

I snapped back to the Students' Union.

'Seriously though' – Buster for once actually did sound serious – 'that could have been fucking, fucking dangerous! You could have both fucking died!'

'Nah! It'd never have happened.' This was handsome Taff piping up for the first time, almost singing his sentences in a Welsh accent. He was two years ahead of me, doing a master's in geology – which was doubly impressive when you knew that he was the first in his extended Welsh family to even make it to university. He and I got on really well.

'I've already seen Peter die,' Taff lilted, 'and he doesn't do it very well. He just doesn't get into the spirit of it. He insists on getting up again.'

It was a story I'd heard him tell before, but I couldn't remember who else around the table had already heard it.

'Was that in France?' checked John.

'Yes,' Taff acknowledged. 'We were in the French Pyrenees on a skiing holiday (although there was hardly any snow, but that's another story) and a few of us decide to go on a trek up a mountain pass. High, high above the valley' – he said 'valley' in a very sexy Welsh way – 'we stop to look at the view. Peter decides to demonstrate a glissade.'

'What the fucking fuck is a "gliss aid" when it's at home?' Buster enquired with his usual intellectual curiosity.

'It's a way of quickly sliding down a snowy slope; I do it in a crouch, balancing on one foot and using it as a ski,' I helpfully interjected.

'Sounds a fucking stupid way to go down a mountain but carry on.' Buster looked back at Taff, who resumed his story.

'So, Peter sets off on a rather nice glissade, with me thinking, "It's a little steep but whatever," and he's picking up speed, and I'm thinking, "It's a little fast but whatever," and then he crosses into an area that looks a little shinier than before, and he suddenly spins around and we all realize that he's sliding on ice, and I'm thinking, "What the *fuck*!"'

There was an appreciative chuckle but everyone was hanging on Taff's every word – even John and me, who knew what was coming.

'He's now accelerating down the mountain slope, head first, totally out of control, and for the first time I realize he's very rapidly approaching a precipice. This is a *serious* cliff edge we're talking about here – hundreds of feet to the valley bottom. And I'm thinking, "Fucky fuck fuck!!!" And *then* I realize I don't have to worry about the precipice because the rock that Peter's head is about to smash into will kill him a few seconds before he sails over the edge.'

'Fucky fuck fuck, indeed!' murmured Buster.

'And a second later,' Taff continued, 'I swear to you I see the top of his head hit the rock square on – this is a proper boulder I'm talking about here, not a glorified pebble – and his head snaps up, and his body is thrown up in the air, and he bounces, and spins through ninety degrees and slides sideways for a couple of seconds as he reaches the cliff edge and, just like in the goddamn movies, scudders to a halt – I *swear* to you – six bloody feet from the edge. And I'm thinking, "Well, that's a bit ironic 'cos he's dead already," and then, fuck me, he stands up and waves!'

With two of his audience incredulous and two of us with

knowing smiles, Taff explained how I'd cheated death. On the day in question, oblivious to the fact that my cranium was supposed to be caved in, I'd registered with relief that I had not shot into space and started the long climb back up the slope to my friends. They in turn had started *down* the slope – which I thought a little strange, but supportive.

I had only been about a quarter way up when we'd met. After expletives, hugs and handshakes, I'd been told to turn around so we could forensically examine where I'd died.

The explanation for my non-death turned out to be nothing but simple physics. And a bit of luck. Invisible until viewed from the side, a small snowdrift had built up along one edge of the boulder in the shape of a perfect ski slope. As my head had approached the rock, it had been flicked up this slope and launched into the air, quickly followed by the rest of me. All I'd felt was that it had got rather bumpy for a while.

'You are *seriously* lucky they didn't have to rebuild you, better than you were before, better, stronger, faster . . .' Nick was attempting an American accent for his Six Million Dollar Man impression.

'We have the fucking technology!' completed Buster.

'Talking of which,' – I'd been meaning to ask this ever since we'd all gone to see *Star Wars* a year earlier – 'am I the only one to wonder how Darth Vader eats, drinks and goes to the loo?'

Yes, was the consensus around the table.

'It's just that it's actually a rather interesting technological challenge when you think about it. He must have been completely replumbed.'

'Well, that's *really* useful to know,' Buster thanked me. 'I'll be sure to bear it in mind if I ever turn into Darth Fucking Vader.'

MND with Attitude

'The point is, I'm going to have the same problems as Darth Vader.'

My allusion appeared to have missed the mark, so I clarified: 'Eating, drinking and going to the loo. These aren't medical problems – they're engineering problems! So they have engineering solutions. Actually, they're very simple solutions. But they're *liberating*.'

That got Tracy's attention.

'What I'm proposing is that we replumb me.'

That *really* got her attention.

'Three operations in one: a tube directly into my stomach for food and drink, the "input"; a tube from my bladder for urine, "output no. 1"; and a tube from my colon for faeces, "output no. 2".' It sounded really simple that way. 'A gastrostomy, a cystostomy and a colostomy,' I added, just in case it had sounded *too* simple.

'Aaaah . . .'

'Of course' – I said 'of course' because I knew Tracy was a nurse by training and therefore knew she would understand – 'I don't just want a bog-standard colostomy because then I'd be left with thirty centimetres of redundant colon that would mean I'd still need to be toileted every few days because of anal mucus discharge. Consequently, I would like us to get rid of that final thirty centimetres, basically just remove the unnecessary piping left in the basement.'

I looked at Tracy as if I'd just laid down a royal flush. Nothing. Again, the poker face. Finally, she spoke:

'Well, that's *very* helpful.' Good start. 'But . . .' Oh! Then she very, very gently broke it to me that hordes of NHS funding commissioners would be picturesquely skating off into the sunset across a newly frozen hell before any surgeon on the planet would actually do what I was asking. After all, my bladder and my colon were fine, and would remain fine. Why would any surgeon damage perfectly healthy organs?

However, to Tracy's eternal credit, she added the words: 'But let's try.'

This left the pesky issue of carrying on breathing after the muscles that inflated my lungs would inevitably give up after a couple of years. I wanted to ensure that my local NHS respiration expert and I were on the same page, so I grabbed the earliest opportunity to meet up with him. I knew he was called Jon, and I knew he had a stellar reputation, so I was looking forward to getting to know him.

As Francis and I were invited into a large consulting room, we noticed that the person introducing himself as Jon was flanked by a couple of colleagues. For a moment, I felt flattered that so many of the respiration team at the hospital should be diverting their attention in my direction. Then I realized that Tracy must have tipped Jon off about me and he'd chosen to bring along reinforcements. I had a horrible suspicion that Tracy was understandably telling everyone that somewhere along the line I had developed a case of 'MND with attitude'.

After initial pleasantries, I decided to save time by sharing with my newest best friends my views on ventilation. I assumed this would be reassuring. All I said was:

'Clearly, based on what Tracy's already told you, you'll understand why I'm going to have a tracheostomy [a ventilator tube inserted through a hole direct into the windpipe] earlier rather than later. However, what you may not realize is that for equally obvious reasons I see my inexorable drift to becoming almost fully locked in as an unparalleled opportunity to conduct some cutting-edge research into cybernetic augmentation.'

Closely monitoring Jon and his team as I was speaking, I concluded that they probably played poker with Tracy. Jon finally spoke for all of them:

'Well, thank you for that. Of course, regarding the trache, there's plenty of time if you change your mind.'

'Aaaaah,' I replied, 'you *really* don't know me yet, do you?'

Half an hour later, Jon and his team were beginning to know me quite well.

'So, let me get this right,' Francis began in his This Is Important voice. 'You're agreeing that if we go down the tracheostomy route then, unless we're unlucky, you can keep him breathing indefinitely?'

'Yes. You're right that a few people suddenly die for reasons we don't understand. And pneumonia can be a problem. But in general, yes, we should be able to keep Peter breathing.'

'In which case, why's he going to die at all?'

Jon shrugged.

'Heart disease? Cancer? Who knows? But if we're lucky, with everything you're planning, it won't be MND.'

I luxuriated in the decadence of reaching for my tall glass of ice-cold beer at eleven in the morning, looked up from the bar table across to the white beach already shimmering in the Caribbean sun, watched for a moment as an unaccompanied dog took himself for a walk along the turquoise water's edge, and I registered that Francis by my side was happily people watching.

My back was still aching dully from the day before, when

my wheelchair had tipped up backwards as I'd overambitiously attempted to negotiate a steep ramp in order to cross the road. Somehow my head hadn't cracked too hard on the pavement. But my spine had.

It seemed that neither the concept of wheelchair accessibility or indeed health and safety had made it as far as the Caribbean. Or maybe there simply weren't any disabled locals. Either way, the fact that I was now 'confined' to a wheelchair meant that it was perfectly possible to be lured down a ramp on the curb to cross a busy road only to find there was no equivalent ramp on the other side to escape to safety.

My lucky though bruising accident the previous day had been a bit of a wake-up call. Just before we'd set off on our latest travels, I'd realized that I needed to inform my travel insurers of my brand-new diagnosis. For the past decade, we'd had year-round insurance from Lloyds to cover all our trips. And I'd banked with Lloyds for over thirty years. I dutifully spoke with an agent for a full hour, explaining that for now at least the only thing wrong with me was that I couldn't walk. Yes, my doctor was very happy for me to travel. No, none of my medical team expected any short-term problems from my MND. Well, yes, technically it was considered a terminal disease, but it wasn't really.

A few days later, just as we were due to leave, I'd received a standard letter refusing me travel insurance. To be accurate, it very kindly offered to cover me as normal *except for* absolutely anything vaguely connected with or attributable via any tenuous link to me having MND. I'd expected at most an unwarranted slight increase in my premium. I could only imagine what a body blow this would feel like for someone newly diagnosed with MND and already struggling to keep their head above water. It seemed cruel and totally unscientific. I was gobsmacked. I was also uninsured.

We'd risked it. There hadn't been time to find a firm to insure me, we didn't know how to find such a firm and we didn't even know if such a firm existed. And after all, there was nothing yet wrong with me other than bad legs. People didn't even need to inform their insurers that they couldn't walk well. My MND was, for now, no reason even to up my insurance risk, let alone refuse me. Having never claimed on travel insurance before, Francis and I had reassured ourselves that it was highly unlikely that I would need to this time. And if I did, it would have nothing to do with MND.

And then my wheelchair tipped back. What if I'd cracked my skull? I could reasonably have argued that my accident was purely the result of being in a wheelchair and could have happened to any able-bodied person, but how could I successfully have argued that the reason I was in a wheelchair was anything other than MND? If I'd needed to be flown to a specialist hospital (inevitably, on the US mainland) and had run up huge medical bills and needed to be repatriated on a medical flight – I wouldn't have been insured. And we'd have been bankrupt.

I banished such thoughts with icy beer. Despite the welcome deep shadow of the bar's interior, I struggled to read the screen of my laptop. It had started off as an email to far-flung friends around the globe to bring them up to date about my diagnosis whilst reassuring them that I was otherwise in fine form. It had ended up as something of a manifesto:

From the outside looking in, the fact I've recently been diagnosed with motor neurone disease might seem rather depressing: Over the next few years my body, but not my brain, will shut down till I potentially can't breathe or – if I use a mechanical ventilator – till I become locked in, unable to move.

But I think that's *completely* the wrong perspective. Instead, look at it from my brain's point of view. Imagine the extraordinary journey that, as an increasingly disembodied intelligence, it is now embarking upon – with me (its self-aware bit) along for the ride.

Unless I'm unlucky, my brain should remain fully functional for the duration, but it's going on a very peculiar and increasingly lonely one-way voyage into a Dark Void that we already know is hostile to life, from where it's incredibly hard to get *any* information back to the Real World, and where the only information *in* (via my eyes and ears, which should keep working) is like a fixed webcam-feed looking at, if I follow the path most others have taken, a rather boring nursing-home bedroom ceiling.

In which case, thank goodness this is the twenty-first century! If my brain and I are locked on course into that same Dark Void that's been swallowing people since – we assume – prehistory, then on this occasion at least let's prepare for a *proper* Voyage of Discovery. Let's be fully *SCIENTIFIC* about this.

I want to take every hi-tech support possible *with me* into the Void; I don't just want to survive in there, I want to *THRIVE*!

Yes, it's a bit rebellious; as ever, my slogan is *BREAK THE RULES!*

This is what my rebellion means. Before I travel too far, I want to set up really reliable life support – breathing and maintaining other bodily functions will largely be mechanical issues, not medical ones. I want great communication systems – both in *and* out of the Void. I want new hi-tech senses and robotic abilities to replace the ones that get cut off – but I want them so that my brain will still be using all its innate processing power, which for any one of us overall still far outpaces the most powerful computers.

And I also want to force some light into the Dark Void, push back the Nothingness and populate it with cyberspace, virtual reality, augmented reality and artificial intelligence – with modern technology there is *no* reason why I must be isolated, lonely or unstimulated.

For instance, as a lifelong writer and music- and art-lover, I want to help push writing and music and art into new realms. In my ultimate straitjacket, I don't just want to be able to stimulate myriad parts of my otherwise starved brain by creating articles and speeches and music and graphic art to express the complex emotions of what it feels like to be marooned in an alien parallel world; I want to write a book about how I travelled there, compose a *Symphony from the Dark Void*, create an artwork entitled *Metamorphosis*.

And send them back to you.

If we're clever enough, my bizarre one-way journey into solitary confinement might eventually end up somewhere remarkably like home. OK, a virtual home, but maybe one that is even more comfortable than the one I was forced to leave, even easier to navigate, even more secure, even more fulfilling.

But above all else, as on every well-conceived scientific voyage, I want to push back the frontiers of knowledge in ways that, if we do it right, hopefully could help millions – even billions – of people.

Some likely spin-offs from the research are pretty obvious, such as learning effective ways to revolutionize the lives of *everyone* paralysed by accident or disease. Yet profound disability and loneliness are often also functions of plain old age; here too there should be widespread benefits.

And there are other, far less obvious, spin-offs. As the sophistication of artificial intelligence continues to explode, we humans need to experiment with how to seamlessly tap into it, how best to use it to amplify our *own* intelligence. Or indeed to compensate for dementia. Otherwise, as a species, we risk being simply left behind.

For all these potential research spin-offs, there's a huge advantage that comes from the fact that the hi-tech support systems I'm going to experiment with are all *computer*-based. Current rates of development mean that a prohibitively expensive piece of kit costing £100,000 today will cost only around £3,000 a decade on.

That makes any spin-offs affordable to the majority. And, to my mind at least, it is the image of *that* glittering prize that ultimately makes my voyage into the Void feel not just bearable but worthwhile.

In terms of impact: I won't be isolated at all.

In terms of a sense of purpose: I'll have one in spades.

What's more, there's a final intriguing thought that keeps flitting through my head as I prepare to set off into the Void on what is for me, quite literally, the experiment of my life: wouldn't it be interesting if, as a result of all these 24/7 hi-tech monitoring and support systems, I eventually ended up living *longer* and with *more* abilities than if I'd never even been diagnosed with motor neurone disease in the first place?

Peter's Second Rule
of the Universe

HUMANS
matter because they
BREAK THE RULES

Taking the Initiative

The responses I received back as a result of my rebellious manifesto were as mixed and extreme as we found our experiences of island-hopping in the Caribbean to be. Some were amazing. Some were frankly appalling. Most people started with expressions of huge sympathy for me. A few registered that Francis deserved sympathy too. But that's where all similarity ended.

Some distant, hardly known acquaintances went out of their way to offer support and make suggestions. Some (I would have sworn) close friends wrote back that my diagnosis made them very uncomfortable – in a tone that suggested that I should have shown better manners than to tell them about it. Some former colleagues sent long replies that did very little other than to rephrase everything that I had written to them.

Some, having briefly acknowledged my carefully prepared document, proceeded to write about something else and never reference it again. This, at least, was something I'd already experienced back in Torquay. Vinny had, by the time of my diagnosis, been one of our closest friends for a couple of decades. He'd listened attentively as Francis and I had broken the news, run through the implications and offered reassurances. Then he'd pulled a face and spoken for the first time:

'I sometimes get pins and needles.'

'Excuse me?'

'Sometimes my hand goes a bit numb.' Francis and

I must have looked confused. He took this as a sign of encouragement. 'And sometimes I wake up suddenly as I'm just falling asleep.'

Francis felt the need to step in at this point.

'Did you hear a *word* of what we just said?'

'Yes, but I'm worried that there's something wrong with *me*!'

This was par for the course with Vinny; possessed of many positive character traits, he had on occasion been known to be a tad self-centred. Just after Francis's mother died, when he had broken the news to Vinny, the immediate response had been:

'I've recently been getting headaches . . .'

In extreme contrast to Vinny were the very small number of friends who offered practical support rather than words. Anthony, still running the opera world from Chicago, put me in touch with contacts in music therapy. Sven, a long-time friend from when he was a management consultant, now a director at Sanofi (coincidentally, the pharmaceutical company that made riluzole, a drug I was taking), organized an analysis of all the current research around the world relevant to what I wanted to do. And Michele, a long-time friend from when *I* was a management consultant, decided to focus on getting me some help. She drove a hundred miles to talk things through.

'We need to get your message out! If you're going to inspire people to conduct the research you think is needed, to change what it means to have extreme disability, people need to hear you.'

'My problem is that all my old contacts are out of date. And none of my old colleagues seem to know who it is I need to get to know now!'

'OK, if *you* don't know who *they* are, we need to find a way to get *them* to contact *you*.'

Way back, even before I'd first met her fifteen years earlier (when I'd been invited in by the director general to analyse the Unwritten Rules of the BBC), Michele had been responsible for sixty hours of news programming a day. So I'd taken for granted that she'd know who to contact.

'I have absolutely no idea who to contact for something like this. Let me think, and I'll get back to you.'

True to her word, she emailed me a few days later suggesting I reach out to a weekly columnist of *The Times Magazine*. Years before, Melanie Reid had had a riding accident and become an instant tetraplegic. She now wrote an award-winning piece called 'Spinal Column'. Michele told me she didn't have a direct route to Melanie but *The Times* would probably give out her email address and it was well worth a shot. Also, she added, she was going to contact someone called Pat who'd worked for her at the BBC and now had a successful TV production company called Sugar Films. Maybe he'd have some ideas. He might even be interested in getting involved. Her final shot was: get on to social media.

I took forever trying to compose what I hoped was a relatively short but compelling email that encapsulated my vision of thriving with MND. I finally got through to *The Times* editorial desk and after a suitably lengthy interrogation they gave me an email address for Ms Reid. I tried it. My email bounced back. Two days later, Michele tracked down an alternative. I tried that. And heard nothing.

A fortnight later, there was a ping on my email and a hugely supportive reply from Melanie. We began a very twenty-first-century conversation: warm, friendly, informative, insightful, stimulating and totally textual. And then, just before my birthday, a little over a month after our dialogue began, *The Times Magazine* of 14 April 2018 published her column about me. After a little context-setting, she wrote:

> Peter Scott-Morgan's attitude is fascinating because he's a robotics scientist, a writer and an expert in organizational systems – those unwritten rules of the game that drive society. A free-thinker who seeks to challenge assumptions (in 2005 he and his partner were the first in England to register as a gay couple with a full marriage ceremony), he intends to embrace his progressive illness and use himself as a guinea pig to demonstrate how much better we could deal with bad stuff.
>
> I'd call this brave. He'd call it rational . . .

She continued spreading the word for a whole page, ending:

> How we need trailblazers to force change at every level.

The message was out. Within a few days, I'd been approached by three TV production companies – including Sugar Films, Michele's friend's firm. Of the three, only Sugar seemed passionate about telling the story that I believed needed telling, rather than a sensationalized human-interest story about how awful it was to have MND and how brave Francis and I were, so I signed with them. Shortly after, they got a commission from Channel 4 (a major UK broadcaster) for a primetime documentary about me and my ideas. When that was aired – in maybe a year or two – the message would *really* be out.

Which was great. But I couldn't possibly wait that long. I needed to leverage the forthcoming documentary to interest a few companies in joining me to conduct some initial research. I also needed some other platforms to get my message out earlier rather than later.

'Why don't you put yourself up for election as a trustee of the MND Association, just as that woman said?'

This was such an unexpected suggestion that for a moment I thought Francis was joking. He was referencing a call I'd just had with one of the local representatives of the Association, who'd encouraged me to become a candidate in the forthcoming elections. My response to her had, at best, been non-committal.

'Well, for a start, none of the five thousand members entitled to vote know who I am, so I wouldn't get in. Second, from what she said, nearly all the available seats on the board will almost certainly be filled by existing board members up for re-election, which means in practice there's only a one-in-five chance of election, so I wouldn't get in. Thirdly, the Association only seems bothered about eventually finding an elusive cure. There's absolutely no mention of technological treatments anywhere on their website. They're irrelevant to what I'm trying to achieve, so I wouldn't *want* to get in. Fourthly –'

'Wait, wait, wait! Just because they don't get it *yet* doesn't mean they don't *need* to get it. Who better than you to educate them about the potential you see in a tech approach? Put yourself up for election on a ticket of cutting-edge research to thrive with MND. If you get on to the board, maybe you'll be able to help the Association and maybe the Association will be able to help you. If, after that, they *still* don't get it, you can resign.'

After further discussion, I capitulated to Francis's logic, drafted a precis of my manifesto, turned it into my candidature submission, threw my hat in the ring, and then finally got around to building a social-media presence.

It was about this time that I received the news that

the colorectal consultant at my local NHS hospital was willing to discuss my ideas about replumbing my insides. A breakthrough. We arranged to meet. A few weeks later, on a lovely spring day, I wheeled myself into his consulting room.

'You must be Peter!' He held his hand out to shake. 'I've heard a lot about you and your ideas.'

I'd done my research on him too. It turned out he was exceptionally well qualified to discuss my plans; he would have been able to hold his own at any teaching hospital in the world.

'Nick! How very lovely to meet you.'

I wanted this relationship to work. More than that, I *needed* him to care about my plans as much as I did, to listen and understand proposals that I knew were baffling to so many of his peers. Instinctively, I tried to elevate it to a scientist-to-scientist discussion.

I stopped talking about me and I started talking about MND, about quality of life, about the holistic view of clinical care, working up to what I hoped was a suitably persuasive ending:

'. . . and therefore, I think we have a wonderful opportunity to pioneer an elective surgery for MND – where previously there has been no choice and no hope – in the form of a tripleostomy.'

I'd made up the name the day before, having decided that my idea warranted an academic-sounding descriptor. At least that would carry slightly more credibility than 'replumbing'.

He asked a few questions for clarification, then started to shake his head. Then broke into a grin.

'It's a no-brainer! Of course the NHS should be offering this. I'll put a team together and we'll operate as soon as possible.'

Although all my ideas about how MND could be treated

had always made perfect sense to me, I was still pleasantly surprised to hear a health professional agree to put them into action. True to his word, he pulled together a team comprising a top anaesthetist, an upper-gastrointestinal surgeon and a urology consultant. This in itself was a major achievement – surgeons from different disciplines rarely even meet in a hospital, let alone actually talk. They devised an overall operation that was cleverer than just combining three smaller operations, and in the process reduced some of the risks (although significant risks remained). It could also all be done via keyhole surgery. So far, so impressive. But it soon became clear that the *really* original bit was the anaesthesia.

We found there was hardly any research data on how best to anaesthetize people with MND – especially for major operations like mine was going to be – for the very simple reason that it hardly ever happens. Doctors instead appear to say, 'Oh, well, conventionally we'd give him a heart bypass but he's going to be dying anyway, so why bother.' And the bypass doesn't get scheduled.

I still find it staggering that, when it came to trying to decide the best way to organize my tripleostomy, my anaesthetist, Maree, had to go back to first principles and rethink her usual options, rather than just look them up. Some of the standard anaesthetic options appeared to risk accelerating MND, so they were ruled out. No muscle relaxants were to be used because, thanks to my MND, I might not respond as expected. However, the big concern was that I might not be able to come off the respirator after the operation was over: I might never be able to breathe on my own again.

Maree sat Francis and me down and began a serious conversation:

'There is the possibility that, far earlier than otherwise, you'll find yourself dependant on a ventilator full-time.'

I'd long ago thought this through, and I was unequivocal. So was Francis.

'I intend to be locked in for very much longer than the time that it takes me to get there. All we need to focus on is my long-term quality of life. How long it takes for me to become locked in isn't really something we should get too hung up on. The worst that is likely to happen after my triple-ostomy is that you can't get me to breathe unaided, in which case you'll whip me back into theatre and give me a trache-ostomy. I'll get into *The Guinness Book of Records* as having had the maximum number of ostomies in one day. It's a win-win.'

Somebody to Love

Looking back, I made the three most important decisions of my life all in the same year. Each choice was massively unlikely to have occurred at all, and each eventually had an equally unlikely but huge impact. But without all three, I'd never have made it to cyborg status. If I weren't a scientist through and through, I'd call it a fateful year. As it is, I can only resort to calling it *massively* unlikely. The year was 1979.

In reverse order, the third most important decision of my life was so unlikely as to almost be random: it was my discovery of a telephone number. During the bleakest days of February, still a virgin, I decided that I really needed to up my game and find my romantic ideal: Avalon. But where might I find him? I felt absolutely no pull to conform with the Unwritten Rules of the gay scene and seek out either of the two gay pubs in London, which I considered a meat market; I wanted romantic love, for life. To find that, surely I needed to spend some time with someone. And be sober.

In desperation, I lifted out the contents of a drawer in my bedroom to reveal an innocuous envelope buried at the bottom. Inside were a few apparently innocuous old newspaper clippings that were actually treasured articles from the only copy of *Gay News* I had ever managed to get hold of (thanks to one of those boozy dinners with a former music master). It came out monthly and was the single lifeline to the gay community in the UK. But these articles were years old. They were also no help. There was no suggestion of where I might find love. Nothing. At all. Except, on the back of a short article, half of an advertisement cut vertically through the middle: 'Th, Cliff H, Hote'. And in smaller letters: 'Torquay', followed by five digits. This had to be a telephone number, surely. It was also my only tenuous link to the gay world. The third most important decision I ever made was to ring it.

'No, I'm sorry, you've got the wrong number. This is a restaurant.'

I apologized and was about to ring off when I suddenly realized that the man on the other end hadn't the slightest idea where I was ringing from or who I was – so I could get away with saying anything. I explained where I'd found the number.

'That must be years out of date, dear. You want Alan's place, the Cliff House Hotel. It's exclusively gay now, you know. Let me get you the number.'

An exclusively gay hotel! I'd never heard of such a thing. And on the coast in Devon too. Nice and far away. I looked in my diary, found there was a long weekend coming up at the end of March, rang the number and spoke to the friendly assistant manager. Yes, they were exclusively gay – the only such hotel in the country. Yes, they could fit me in for three nights that weekend. Yes, he was sure I'd love it; in fact, he'd show me around personally. Yes, that was a promise . . .

The Cliff House was a sprawling crystal-white Victorian villa glimmering against the blue sea of Tor Bay. As I walked down the drive towards the double doors into its entrance hall, the setting looked perfect. With a smile on my face and a spring in my step, I entered.

There was no one inside. I waited, the ceiling soaring overhead. The décor, I noted, was rather green. I called out. An old man in his fifties appeared, dressed in a similar colour scheme. He turned out to be the proprietor of the hotel, Alan. No, the assistant manager wasn't there at the moment but would surely show me around later if that's what he'd promised. Yes, he'd point him out to me when he returned. No, most other guests were yet to arrive. Yes, they'd be full for all my stay. No, he didn't think there'd be anyone my age.

At that moment, four giggling men emerged from the huge lounge I could see to the right of me. They were ancient and rather effeminate. Alan immediately joined in the joke – which was impressive, given that he couldn't possibly have known what they found so funny – and wished them well on their way as, still giggling like a gaggle of schoolgirls, they went out to explore the harbour. What had I been thinking? All of a sudden I felt I'd made a terrible mistake.

My upstairs room was in what once must have been a long stable block. It had a large window overlooking a courtyard, and it was at this window that I was standing twenty minutes later, bemoaning my lot. I'd booked into an old people's home! And here I was, stuck till Monday. It was only Friday. Worse, the assistant manager who had promised to show me around was probably as geriatric as everyone else. I'd be stuck with him too. And then I saw Avalon.

He was striding down a long flight of stone steps that led from an almost hidden door in the high wall at street level down to the courtyard below. It looked like the secret exit from a castle. And he looked like a young lord. Indeed, with his flowing, shoulder-length, red-gold hair, he looked exactly like *my* young lord, just before his father was killed and Avalon became a prince. He was slim and athletic. And, as far as I could tell from this distance, around my own age. Then he was gone, hidden by the corner of the main house.

It began to drizzle. As drops gradually congregated on the windowpanes and then dribbled downwards, I continued to stand by the window looking out. I felt confused. No, in truth, I dared to feel hope. Maybe Avalon was a guest. Maybe I could get to know him. Maybe . . .

But then the analytical side of my brain kicked in. Surely a guest wouldn't know about the secret entrance. And Avalon had been striding in as if on a mission, without looking around. Maybe he was delivering or collecting something. Maybe he had no direct connection to the hotel. Maybe he was straight. Typical.

In front of me, on the ground floor directly across the courtyard, I could see through a couple of large windows into what appeared to be the hotel kitchen. I hadn't noticed any movement in there since I first looked out of my window. And then there he was again – even with his back to me, unmistakably Avalon. The hair. The slim athletic figure.

For reasons that made little sense, I felt my heart instantly racing. I felt excited. I felt happy. And I didn't care that it made little sense. Then he turned, a kettle in his hand, and walked towards me to the sink in front of one of the windows, turned on a tap and waited, and looked up.

For a second, I thought he was looking at me but, before I could smile at him, I realized he was staring slightly to one side and hadn't seen me. I froze, and like a voyeur I watched him. He was gorgeous. It was the first time I'd really seen his face. It only took a few seconds to fill the kettle, but in that time I fell in love. Then he looked down, turned the tap off, turned his back and walked out of sight.

What was I to do? I waited to catch sight of him again. After five minutes, he still hadn't reappeared. But surely he must still be in the hotel? I must try and find him. But not like this; I was still in my travelling clothes. I went to the small wardrobe where I'd just hung the contents of my suitcase and after due deliberation selected a suitably eye-catching ensemble: exceptionally tight red trousers, a torso-hugging brilliant-white T-shirt, topped off with a silver blouson jacket that looked like something an astronaut would wear in the twenty-second century. Bearing in mind this was the late seventies, I looked *awesome*.

I slowly walked down the long hall from the courtyard to the lounge, accompanied by the hotel sound system, which had begun playing the soulful opening chords of Art Garfunkel's 'Bright Eyes' as soon as I'd entered. It was my favourite track at the time. A good omen? I looked into the dining room, the TV room, reception and the entrance hall in case he was there. He wasn't. That left the enormous lounge. I looked through the doorway. And saw nobody. I went in – there he was.

He was sitting across a small table from Alan, in a far corner by some French doors, with his back to me, his red-gold hair cascading to his shoulders, teapot and cups in front of him, a golden Labrador asleep at his feet. The music was louder in here.

Alan saw me, got up, walked over and (rather conspiratorially) pointed at Lord Avalon and mouthed: 'This is our assistant manager, Francis.' He then stood back expectantly, as if waiting for a show to begin. I stood there, smiling my thanks to him, every fibre of my being straining to put on the best performance of my life. I walked towards the personification of all the romantic hopes and dreams of my teenage lifetime, and the curtain went up.

Not too close – you don't want to startle him. Not too loud. Calm. Friendly. Warm. Loving.

'Hello! I've come to hold you to your promise . . .'

Maybe I'd subconsciously delayed my utterance to fit with the music. Maybe it was coincidence. But as he turned around to face me, the music swelled into the powerful final chorus of 'Bright Eyes'. And, sure enough, those were what hit me first. They were piercing blue. And from the moment they locked on to mine, they never looked away. Then he smiled, and part of me registered that I'd never fully imagined what Avalon looked like when he smiled at Rahylan.

Alan explained who I was, Francis remembered me, and for the rest of the day he made good on his promise. It was one of the few times in my life when I felt like an actor in a preordained play – everything simply worked out right. Unlike me, Francis had left school at sixteen (when his mother said he was 'tall enough') but, like me, he had ambitious dreams. Unlike me, Francis was worldly-wise and street-smart (having escaped from a poor, working-class upbringing) but, like me, he was proudly gay. Unlike me, Francis was anything but a virgin (despite being only two years older) but, like me, he was looking for true love. What more was there for me to care about?

Less than an hour after we first met, he asked me: 'Do you believe in love at first sight?'

The answer was already self-evident to me: 'Definitely!'

Seconds later he quietly reached out, grasped me by the waist, pulled me to him and kissed me, and kept kissing me . . .

Over the next three days, I rapidly began to compensate for my teenage years of isolation and virginity. I equally hungrily devoured any and every morsel of information about the person I'd fallen in love with.

At our cores, we appeared very similar – gay, ambitious, James Bond fans, atheistic, searching for love. Apart from that, we had absolutely nothing in common. For a start, I was surprised to learn that Francis had never studied Latin at school. Not even for a term. Not at all. I was even more surprised to learn that none of his friends had either. In his world, evidently, this was considered normal.

It was a huge eye-opener to me. Francis was clearly very bright – he felt like an intellectual peer whenever we discussed a subject that we both knew about. But, as we lovingly shared everything about ourselves, it dawned on me that this was the very first person I'd got to know in my entire life who was simultaneously highly intelligent but had hardly any academic training at all. And when I'd explained that I'd really enjoyed fencing, he'd assumed I built fences.

On many subjects that I took for granted, he had huge gaps that he happily looked to me to fill. But it brought me face to face with just how ignorant and uneducated I was in ways I hadn't figured. Wonderfully (for some reason I never felt disconcerted by it), Francis had a deep knowledge of a broad range of non-academic subjects – not least Living in the Real World – of which I knew almost nothing. I'd been cloistered in ways I'd never even realized. And perhaps because of that I found these new fields of study gloriously exotic. And liberating.

Equally liberated, I soon discovered, were most of the guests filling the hotel for the long weekend. Ancient they might have been (most were old enough to be my father, the rest my grandfather), but they certainly had voluble views about me.

'Mmm! A public schoolboy, how erotic!' They always seemed to talk *about* me rather than *to* me. 'The only problem with public schools is that they turn out incredibly well-educated little shits.'

I had to admit that they might well be correct on this point.

'Ooh!' yodelled a slightly mincing queen who, if I hadn't known better, I would have sworn was wearing mascara.

'She's got looks *and* brains, darling!' He was facing Francis. But he was gesticulating towards me. 'I'd hang on to this one, dear,' he recommended conspiratorially.

'I intend to!'

After three heavenly days of falling deeper and deeper in love, it was time to return to my old life. The extraordinary pain of leaving my soulmate was made bearable to me by my irrevocable promise to Francis that I would return. To live with him. Soon. To be precise, on my twenty-first birthday. Which was only three weeks away. This only left the minor obstacles of coming out to my parents and finding a way to persuade Imperial College to let me postpone my under-graduate degree indefinitely.

I took the train back to Paddington, then the District Line to Wimbledon, then marched up the hill to the Ridgway, suitcase held high, until a mile from the station I reached my parents' flat (where I'd arranged to spend the next week). As arranged, I rang Francis as soon as I got in, using the phone in my parents' bedroom for privacy. We talked for an hour. But soon after Ma had put her head around the door and mouthed that Da had just got back from work, it seemed like I must ring off.

It was remarkably difficult. We both finally said, 'Bye . . .' but neither hung up.

Francis broke the silence: 'Are you still there?'

I answered for the rest of my life:

'Always! I'm yours forever.'

Fork in the Road

'How do ordinary people cope?' Francis enquired out of the blue as he parked our Range Rover in a perfectly located disabled-parking bay right on the seafront by Torquay Harbour. Despite the near-telepathic link we'd honed over the decades, this was *such* an open-ended question that I didn't quite know how to respond.

'Huh?'

'Simply getting a blue badge!' He waggled my disabled-parking permit before placing it in view on the dashboard. 'Most people couldn't have convinced them like you did. They've really clamped down on giving permits out. It's almost impossible to get one now. I felt intimidated just trying to read the form, let alone filling it out. How many pages was it?'

'Six, I think.'

'Exactly! And how many closely typed pages did you send them?'

'Twenty-ish.' It had been my first form of many – triggered by a genetics consultant throwing me somewhat by casually describing me as 'already severely disabled'.

'That's my point! You're used to holding things in your head, structuring them, writing them down in a way that's hard to dispute. I couldn't do that. I bet most people who really *are* eligible for a blue badge struggle to present the evidence needed to get one. Without you, I'd be one of them.'

He got out, walked around the back, extracted the active-user wheelchair that we'd bought at great expense (albeit with a token contribution from the NHS), brought it to the passenger door and helped me transfer into it. Even this early in the day – before the grockles, no doubt now finishing their full Englishes in their hotels, guesthouses and B&Bs, flooded the resort – the sun glinting on the bay was bright, the light sea breeze was warm and the palm trees were susurrating.

'Why don't we head to the pier first?' I suggested. It was always either pier then harbour or vice versa. A couple of summers ago, we'd sometimes completed our circular tour within twenty minutes; these days, I was so slow that it typically took an hour. 'We may see some more barrel jellyfish.' These were the size of a dustbin and in the last few years had got into the habit of spending their holidays lazing about around Torquay.

'And what about getting PIP?' Francis persisted, referring to the new type of disability benefit. 'How long was *that* form?'

'Yeah, that was horrendous, about twenty pages!'

'And how long was your submission?'

'Well over forty.'

We'd crossed Princess Gardens and now turned right along the prom.

'Exactly. Most people aren't able to reel off small books like that.'

'I have to say, it was actually one of the hardest documents I've ever had to write. I answered all the questions on the assumption that the whole system was designed to make it as tough as possible to be eligible. So I provided every single bit of evidence I could persuade the medical profession to corroborate. But it *was* a bloody pain.'

'*And* a lot of people who are in exactly the same boat as you are wouldn't have had the time or energy or writing ability to pull it off. It's scary.' He paused for a minute as we navigated around a solitary old man with two walking sticks slowly making his way along the prom. 'And we're *very* lucky.'

'Very lucky.' Unrehearsed and unplanned, my half-baked ideas came tumbling out. 'You know, I've been thinking. I know it's always been us against the world. And we've always been a bit of an island. We haven't needed them. And they haven't wanted us. Well, I think that now maybe they *need* us . . .'

We'd reached the beginning of the Victorian pier, across from the Princess Theatre, and we now turned left on to its planked surface, the sea visible through the cracks.

'What are you talking about?'

'My point is that – and I *know* this sounds stupidly pretentious – I think we actually have a chance to do more good in the next couple of years than we've ever done in our entire lives.'

'Look! There! By the blue boat.'

Two almost transparent giant jellyfish were floating just below the water. I paused in my exposition for five minutes' jellyfish watching.

'So, suddenly you want to become a charity worker?'

'No! Not at all.'

'Aah, so we'll make money from it – that sounds better.'

'Well, no, but it's *really* important. I think that you and I can make a difference. Change things. Change the world.'

'Or we could enjoy the last few summers in which you're relatively well! We'll *never* get this time back again. I don't want you wasting all your energy and *our* time on some grand gesture that you get carried away with and ends up dominating our lives and that *no one* is going to thank you for!'

Even though I didn't fully know what it was I was asking

117

for, I couldn't help but feel disappointed that I wasn't going to get it.

'I know. It was probably stupid. It's just that it feels like our wedding, you know, when we were going to do it all low key and then we realized that a lot of people depended on us, strangers not as lucky as us, not as strong as us. And we were in a position to say "fuck you" to the status quo, so we did, and we were really proud that we did. This just feels the same, that's all.'

We'd reached the steps that led down to the second section of the pier where, as usual, a few men (always only men) stood with their fishing rods, staring vacantly at the thin lines trailing outside the harbour wall. All our lives together, we'd always walked on till we touched the harbour light at the very end. Those days were over. Francis reached down and touched my hand.

'Look, I'm not saying we shouldn't do it, I just don't want us to leap into something without thinking it all through first. If we're going to do it, we've got to do it right.' Suddenly, I knew he was going to agree. Even though neither he nor I yet knew what we were committing to. He stood full height again. 'So, what's your grand plan for how we save the world?'

I especially loved him when he was in this sort of mood: courageous despite caution, brave without bravado.

'You remember how in 1984, in the last chapter of my book *The Robotics Revolution*, I argued that, within our lifetime, we'd reach a fork in the road? And which route humanity chose would change everything?'

'Yes, and they tried to bully you to take the chapter out.'

'And, thank goodness, failed. In one future, the most likely future, the default future, robotic intelligence develops independently, pure AI getting smarter and smarter with humans eventually relegated to the status of pets or pests.'

'Which is the rubbish future that Hollywood always shows.'

'Precisely so! The alternative future that I argued for was one where we *melded* with AI, where we enhanced ourselves in ways that meant we could do things that neither we nor AI could do alone, a form of human-AI collaboration.'

'Which is exactly what you're now pointing out could be helpful to those with MND and extreme disability and the elderly.'

'Yes! Because we've reached the fork in the road. It's now! The general public doesn't realize it. Politicians don't realize it. But the IT industry does. It's just keeping quiet about it. It doesn't want to draw attention to what's going on.'

'And what *is* going on?'

'We're charging down the independent AI route! We haven't all discussed it, let alone agreed to it, we're just doing it! Practically no one has even noticed that there's an alternative route, an alternative future that we're shooting past and very soon – like missing a motorway exit – we'll never be able to reverse back and take it. And we'll never have that choice again.'

'Oh!'

'The alternative future is already becoming less likely. But I think I see how to nudge the future in a direction that would at least offer all of us the option of choosing collaborative AI rather than independent AI, if that's what we want.'

'It all sounds very Hollywood blockbuster!'

'It is – you couldn't make it up. The point is, the ideas I've been working on for the last year could be a *perfect* focus for researching and showcasing collaborative AI *for everyone*. Think about it – who could argue against the goal of helping millions of people with extreme disability to thrive, billions of elderly people to thrive?'

'No one. Well, no one apart from the don't-play-God brigade.'

'It's a bit like JFK's audacious vision of landing on the moon within the decade. MND's a brilliant research challenge not because it's easy but because it's about as tough as it gets. And because of my background, I'm uniquely positioned to be a human guinea pig. I think some really top brains, top companies, might be interested in becoming involved. And that would change *everything*.'

'I get that, but why does it need to be you and me who have to sacrifice what little precious time we may have together? Why can't we leave it to all those people who are being paid a salary to worry about these things?'

'Because I've a horrible suspicion that there's no one else out there in a position to do what we can. Not least because no one else out there – and certainly no one with MND – is even *trying* to do what I'm proposing. On the contrary. The status quo is immensely strong. The default future that we need to be striving to change has already largely been written *with the wrong ending*. But we have the chance to create an alternative. A chance to demonstrate a different way forward, a non-threatening way, a safer way.'

A luxurious sailing yacht was slowly pulling away from its berth to navigate towards the outer harbour, its enormous white sail gradually inching up its towering mast. We watched as the busy crew drifted by before Francis turned back to me.

'We'd better get going.' He smiled. 'An alternative future isn't going to create itself!' I suddenly felt elated. We headed back along the pier, keeping pace with the yacht until it turned to starboard out of the marina and we turned left towards the beach. 'I want you to explain exactly what this alternative future is going to look like. For us. Forget about the world for a moment – I know that more and more will follow. But I want to know what it's going to be like for *us*. I want to understand *our* alternative future.'

We continued along the prom, large Canarian date palms on our right, morning-blue sea on our left.

'Well, as you know, I'm in effect going to turn into a cyborg.'

'Really? I know you've been saying this since we first met. But it's going to be a huge turn-off to most people if that's what you tell them you want to do. You love the idea. But to most people it's scary, even creepy. It's science fiction.'

'Yes, but it's true! If we manage to do even just a little of what I know is possible, I'll become the first full cyborg in human history.'

'But didn't you tell me that people with implants were calling themselves cyborgs ages ago?'

'Yes, I know, depending how we define it, even someone with a pacemaker can be a "cybernetic organism". But I'm going the full monty. Almost everything about me is going to be irreversibly changed – body *and* brain.'

'Aah, the Lounge is just opening. Fancy a coffee?'

We crossed the road, made our way around the outside tables and went through one of the sets of double doors. The long room had a New York loft feel, with damaged walls, exposed air-conditioning and eclectic collections of pictures, mirrors and furniture. Only two of the unmatching tables were yet occupied, so we chose our favourite, in the centre, by the picture windows overlooking the bay. A suitably hipster-looking waiter – young, slim, lumberjack beard – took our order for two double-shot lattes, and I continued.

'It goes without saying that all my physical interaction with the world will become robotic. And naturally, my existing five senses are going to be enhanced.'

'Naturally!'

'But far more importantly, part of my brain and *all* of my external persona will soon become electronic – *totally* synthetic.'

'You're losing me.'

'I mean that I won't just be "old me" any more, Peter 1.0 if you like. The new me, Peter 2.0 if you like, will be most of my old brain (even chunks of that, everything to do with movement, will eventually shut down) *plus* lots of cybernetic augmentations. But my body, apart from my eyes, will be there only to keep my brain going.'

'So, basically, you'll be a brain on a slab in a lab – just like you always said!'

'Well, certainly for a lot of the time I'll interact with the world like that, remotely, via the internet, by remote-control robot, or something else.'

'But I want to interact with *you*! I don't care how you interact with the world. I want to interact with the man I'm married to, not a robot.'

'Let me rephrase that. I guess what I'm saying is that in the future I'm going to *be* partly a robot. That will be the *real* me. I'll be part hardware/part wetware, part digital/part analogue. That's the only way I can think of that will keep me being me. If I'm to keep chatting, joking, keep smiling, pulling faces, keep my personality, keep being *me* – then I have to change, sweetheart. You'll only be able to interact with the robot body and brain on the outside, but you'll know that the old me is still there deep inside.'

Fortunately, at this philosophically profound and disconcertingly emotional moment, our lattes arrived in tall glasses with little handles and, no doubt, a backstory about where the beans had come from if only we'd asked. Thanking the waiter, stirring with the long spoons provided and sipping the foam at the top gave us both the time I suddenly realized that we needed to assimilate the full enormity of what I had just verbalized – and indeed just clarified in my own mind – for the first time.

'OK,' Francis eventually rebooted the conversation. 'You need to tell me *exactly* what you have in mind.'

Unwritten Futures

'You remember how that director at the BBC told me that Maggie Thatcher used to do TV interviews wearing an earphone connected to a team that fed her statistics, clever answers and sound bites?'

'Yes, in her right ear, away from the camera.'

'Well, I want my whole life to be like that! I'm envisioning an AI system that can listen to what's going on and then prompt me, like a satnav, with, say, three alternative suggestions according to different criteria. And of course the synthesizer will sound just like me, my voice.'

'Oh! I'd assumed you'd sound like Stephen Hawking.'

'You know, Stephen Hawking could have had a far better voice ages ago, but everyone recognized his earliest version so he didn't want to improve it. But while my voice is still strong, I plan to voice bank to the highest quality possible – in other words, record in a sound studio every possible combination of sounds – so that we future-proof my voice to take full advantage of all the advances over the next few decades.'

'But I still don't understand how you'll be able to get your AI to say *new* stuff, things that have to come from biological you.'

'What I have in mind is that I'll kick off the initial response I want my voice synthesizer to use while I'm using my eyes to type a –'

'Wait! What do you mean "using my eyes to type"?'

'It's simply that eye-tracking technology I told you about.

My eyes should never get paralysed, so I can look at a keyboard and a computer will work out which key I'm looking at. It uses a clever bit of kit that tracks my eyes.'

'OK, got it.'

'So, having kicked off my voice synthesizer to say a phrase proposed by my AI, I then use eye tracking to spell out a customized insert to slot in after the space-filler – using a hugely predictive text engine –'

'Like on my mobile?'

'Yes, but much better, and customized to where I am and what's going on and personalized to my style. And I also imagine the AI system will be clever enough to judge which of my synthesized emotional voices I should use for different words – conversational, passionate, intimate, and so on.'

'But wait a moment! You're telling me that what you say to me in the future, what "Peter 2.0" says to me, will often be the *AI* speaking, not *you*?'

'It has to be. It's the only way I can have any hope of being spontaneous. Otherwise, you'd have to wait for maybe a minute before I could type out a few words with my eyes. It's *very* slow using even the very best of the hi-tech solutions available now.'

'So I'll never know if it's really you talking to me.'

'It will *always* be me. OK, sometimes I'll be a bit like a movie director nudging an improv actor. But it'll still be *my* movie.' I wanted to add: 'When you hear the words "I love you", you'll know I'm still in there,' but I suddenly began to tear up.

I'd become aware that this was happening more frequently these days, much as it had when my parents had died within three months of each other. Then, and now, I knew a large part of it was due to stress. But I was also on the alert for something called emotional lability: uncontrollable and inappropriate emotions such as laughing at something sad. Sometimes, MND could bring on a form of dementia, including lability.

Fortunately, so far, my higher mental functions appeared to be unaffected and, although poignant moments in a movie could now make me cry, they'd *always* brought tears to my eyes. And at least they didn't make me laugh. I hid the tears pricking my eyes by looking down at my latte, and by the time I'd taken another sip I felt just about able to look Francis in the face again and carry on.

'Of course, what my AI ends up saying may not be the response, with the expressiveness, even the exact same idea, that I would have said unaided. I realize that. But it's *OK*. Especially if it makes me look cleverer, or funnier or simply less forgetful than I was before. The joy is, I won't be faking *anything*. As I keep saying, this will be the *real* Peter 2.0.'

I made it sound jokey and almost incidental, but I knew that Francis's reaction to what I was explaining now was in fact deeply serious and all-important. This wasn't just about me. It was about our relationship. Our future. And the unfair asymmetry of Peter taking it as his right to try to upgrade to Peter 2.0 while Francis, what, just grew older? All our adult lives we had been a united front, a seamless entity, a loving couple who had become one. In my grandiose exuberance, I now risked destabilizing the bedrock of our existence. Francis had fallen in love with flesh and blood, not a cyborg. He might prefer to keep it that way.

'Cleverer sounds good. Funnier sounds *very* good . . .'

'In place of verbal spontaneity, think intelligence amplifier!'

'I like the humour amplifier idea.'

We seemed to have traversed the first hurdle with ease. This was good because, by my reckoning, there were at least another six hurdles to get over. Seven, if there was time and headroom to take Francis through my 'change the world' idea.

'Then you'll love the second idea. It's all about me being able to laugh and smile and express body language and retain my personality and humanness. I want to build an avatar.'

'Like in the film *Avatar*?'

'Exactly! Except it would look like me. Well, me three years ago, before I lost any muscle. Well, my head anyway. And I'm not thinking of something that looks like a cartoon, I'm thinking of something that every instinct tells you is a real human being – *me*. I'm thinking of hardware and software that can deliver that in real time. And if initially we can't quite achieve photorealism in real time, then at least from the very beginning we can achieve it for pre-prepared sequences.'

'You mean like if you were giving a speech or something?'

'Exactly! And the rest of the time my avatar will be lower resolution but able to operate in real time.'

'I *like* that! I love the idea of you still smiling your smile and your face still moving when you talk. You could have a screen on your chest with your full-sized avatar head on it, like a T-shirt with someone's face on it, but yours will move!'

'That's brilliant!' It was. I'd previously been imagining a screen that partially blocked my face; Francis's idea was far more user-friendly.

'It'll eventually be possible to get a really good match between my synthetic voice and natural-looking facial movements on the avatar. But I also want AI-generated facial body language – emotions – based not only on the AI listening to ongoing conversations *and* things like sudden noises but also on it *watching* what is going on, detecting and interpreting movements, recognizing people, and so on.'

'Won't people be upset if they know your AI is listening and watching all the time?'

'What's the difference between that and when people use Alexa or Google Home or Siri? *They're* always listening. That's the way AI's going. More and more devices will do the same. At least the reason my system will be listening and watching everything is because otherwise I won't be able to be spontaneous at all.'

'If you ever get it to quote back to me what I actually said several days earlier, I'll turn you off!'

'I promise I won't! But just imagine how amazing me having an avatar will be. I could give a keynote speech where my avatar is shown on the auditorium screen (not an image of my almost paralysed, speechless body), or do a Skype call or a podcast where my Peter 2.0 avatar is all that people see, or even hold a face-to-face conversation with someone who ends up interacting with my Peter 2.0 *avatar* software, *not* my original wetware at all.'

'And it's actually rather reassuring. We know you're going to lose your ability to speak. We know you'll eventually lose your ability to express emotion, except, I guess, with your eyes. But in reality, you won't!'

'Nobody's ever done this before. In, what, a couple of years I'll have crossed an invisible frontier; the only window you'll have into who I really am will be completely artificial.'

Francis started laughing. Then he explained:

'I've just realized, as I'm looking at you now, I'm looking at a prototype!'

'Yes, and unlike the original, my avatar will never age.'

'This gets better and better . . . Oh! Have you thought of this? What happens if you die? What happens to the avatar?'

I had indeed thought about this. A lot. There were some important implications I had to work through. In fact, totally unexpectedly, I felt I was close to seeing a way around the Uploading Problem that I'd explained to Anthony more than forty years earlier on that Greyhound bus. But for now, I made light of the scenario.

'My prediction is that we eventually get the AI on all this *so* good that one day I'll unexpectedly die – and *no one will notice*. After a couple of days, you may ask me, 'Can you smell something?' and my avatar will answer: 'Can't smell a thing.' Which, technically, will be perfectly truthful . . .'

We paid for our coffees and went back outside.

'You know, if you ever eventually die . . .' Francis began again. 'I mean, if you ever die and I happen to still be alive –'

'That would be *awful*!' I'd always planned to save Francis from that horrific eventuality by living a few painful days longer than him.

'Yes, but what I'm saying is that I'm beginning to realize that still at least having your avatar to talk to, to remind me of things, to program the TV, would actually be a huge comfort.'

'Are you sure?' I was delighted. And hugely relieved. I'd recently been giving a load of consideration to what would happen – OK, let's be honest, what might just conceivably, maybe, if I was unrealistically lucky, *possibly* happen – if I lived long enough for my AI-driven avatar to get as good as I knew it could and *then* my biological body gave up. If that mind-blowing scenario ever came true, Francis's wishes would be paramount. 'Are you sure you'd still want me around after I'd gone?'

'It would still be awful losing you, or at least losing part of you. But having you die but not die would be infinitely *less* awful than being *totally* without you. Just so you know, that's all.'

My first thought was that Francis was blissfully unaware

of the ethical minefield he had just neatly stepped through unscathed. Which was immediately followed by the realization that he had known exactly the significance of his words, which was why he'd casually dropped into our conversation his authorization for me to break a whole host of ancient taboos by attempting to fuzzy the jurisdiction of Death. I realized we had better seal our momentous pact:

'In which case, I'll do everything I can to become one of the Undead.'

'Please do.'

The front was slightly busier, with a few early risers shuffling to the beach in their flip-flops. We crossed over to the prom and turned left, retracing our earlier route from the pier. The bay always looked totally different in the other direction.

'Now, combine all that with advanced virtual reality.'

'Oh, do we have to? I thought that was it!'

'There's *lots* more. It's really exciting!' As the day was warming up, so was I. 'But think about it. Even just combining these first three things – controlling my voice, my avatar and now virtual reality – puts a huge drain on my ability to control everything via eye movement.'

'Is eye tracking really the only way you'll be able to control things?' Francis broke in.

'One day, direct brain-computer interfaces will become very fast. But that's probably more than a decade away. For the moment, eye tracking is as fast as it gets. Stephen Hawking actually used a cheek muscle, which was even slower. But I want to experiment with trying to control a lot more than just speech, more than has ever been attempted before. And that's the problem. Even with what I've run through already, I'll be trying to somehow control speech *and* emotion *and* body language *and* now movement (in virtual reality) – all at the same time.'

'It's not doable?'

'It is and it isn't. My plan is to use the same trick for my AI as my biological brain does: delegation. I'll use such smart AI that at a conscious level I only need to kick off very high-level commands (such as 'move over there') and it just happens without further intervention from me. Exactly like you're walking on autopilot now.'

'Or how you tend to speak without thinking . . .'

'Well, if you think about it, we all speak without thinking of which muscles to move. But when I'm Peter 2.0, delegating as much as possible to AI will give me the headspace to teleport into and participate in a virtual meeting anywhere on the planet.'

'That'll save on hotel bills.'

'Or, far cooler, the meeting could be on any planet in the virtual universe!'

'I get that. But I don't see how you can delegate movement. How can you tell your avatar to make a complex movement without actually showing it what to do?'

'I plan to use the same trick as before. Similar to the three satnav-like options I'll have for verbal spontaneity, I envision being offered physical equivalents – predictions of likely movements I may want to make in VR based on everything that is going on.'

By now, we'd passed the pier and reached the Banjo, a large circular area with a few steps down to a glorious walkway close to the sea that made you feel like you were on the prow of an ocean liner. Even before the Banjo had been restored, Francis and I had always taken the detour to the right, down the steps. This time, for the first time, we turned left. Without discussion. It was yet another door that had silently closed.

Fortunately, the French market was in town and the

stallholders were setting up, providing a temporary entice-ment to drag my eyes away from the sea I loved. Focus, I chastised myself, on those things you still *can* do, not on the growing number that you will never do again.

We skirted the skeletons of the cheese, cuddly toy, fossil and craft stalls, and after a long half-circle we found ourselves back by the water's edge, close to the ocean-going speedboats moored along this part of the marina. We turned left, saying nothing as we both looked longingly at the sleek boats. We'd been planning to buy one since we first met, but a year ago we'd *really* been going to buy one. At last. Only after fran-tically exploring ever wilder schemes to keep our dream alive had we finally, very recently, let it die. Focus on what you *can* do. To Francis, and to myself, I needed to demonstrate the hope and positivity that I genuinely felt, which I knew was the only way to carry us both through what lay ahead.

We passed the harbour master's office and began to smell the salty, fishy lobster pots and nets newly stored for the day by the local fishermen. I'd always loved that evocative smell of a working harbour. Once I was on invasive ventilation, I wouldn't be able to smell any more . . . Focus. Listen to the seagulls. I'll still be able to hear seagulls.

Struggling against the tide of emotion, I pulled myself together.

'You remember those rides in Disney World? That's the feeling that eventually should be possible for me to experi-ence every day. I'm visualizing VR that feels closer to being inside a flight simulator or on a theme-park ride than merely playing a domestic computer game – even David's.'

'Whose?'

'Nephew David. Those games he brought round . . .'

'Aah! They were *amazing*! The shark was incredible.'

'And flying over Paris . . . After all, I'll be strapped into an incredibly hi-tech wheelchair. If we want to simulate G-forces, we can tilt me. And when I'm in VR, only the flesh of the lower half of my face will be exposed. So it'll be easy to create sensory effects such as warmth or breeze. And in the process push the boundaries of VR. So maybe some of the game producers will be interested in getting involved. One day, I want to create a virtual multiverse where anyone – with or without extreme disability – can live on an equal footing.' I paused, remembering my teenage dream. 'You and I could finally explore Salania together . . .'

'You've been promising me that since we first met and you told me I was the spitting image of Avalon!'

We'd almost reached the Old Harbour, but a small group was blocking our view as they crowded around the railing, pointing at something in the water directly below, holding their phones out to film it. I looked up at Francis.

'I have a simple dream: one day, you and I will walk across a high mountain plateau in Salania, our phoenixes calling overhead, ahead an incredible view straight out of a Maxfield Parrish painting, and we'll get to the very edge of the towering precipice and stand hand in hand looking out on an almost unbearably beautiful alien landscape in a galaxy far, far away. And then we'll effortlessly fly to the next peak to watch twin suns rising out of a turquoise ocean into an impossibly perfect sunrise. And in those moments, we will be free . . .'

He said nothing. Simply smiled a tired smile, and then kissed me.

'What are they looking at?' I eventually asked. He left my side for a moment to have a look, eased his way through the throng to the rail, leaned over for only a few seconds, then pulled back and returned to me.

'It's only a couple of barrel jellyfish. They're right against the harbour wall, so you wouldn't be able to see them. We've seen them before!'

I knew Francis could spend all day watching barrel jellyfish. So could I. Without further discussion, we circumnavigated the ogling crowd, passed the drying nets and crossed to the other side of the harbour over the incongruous but beautiful Millennium Bridge, its stainless-steel towers glowing in the morning sun. The Brixham ferry was surging into view, about to throttle back as it reached the outer harbour.

'Most people use VR as an opportunity to *escape* their lives.' I quickly glanced back towards the lucky group still enthralled by what they alone could see, squinting my eyes to catch a glimpse of anything in the water, but the light was in the wrong direction. I flicked my head back in Francis's direction. He hadn't noticed. He was resolutely looking forward.

'I want to use VR to *reclaim* ours.'

More Life Choices

I'd set the hi-fi in my old bedroom to keep playing the same Art Garfunkel 45 continuously. Three minutes and fifty-seven seconds after the single started, the needle lifted up, swung back to the beginning, crackled and thumped on to the outer edge of the disc, and the oboe of the introductory chords of 'Bright Eyes' would start again.

The record, accompanied by a gorgeous picture of Francis, had arrived on my first morning back. He had bought it and posted it straight after waving me goodbye to the frantic accompaniment of the guard at Torquay Station blowing his whistle – desperate to remove as quickly as possible at least one of the two miscreants who had flagrantly flouted common decency by kissing through the train-door window as the locomotive juddered into motion.

Francis's photo was now blatantly propped against the record deck. I had explained to my mother that the record was a gift from the sultry-looking person with the golden-haired chest in the picture. I played the record continuously. At a reasonably high volume. I spent one or two hours *every day* talking – as I made clear each time – to the same young man in the picture. Surely I was being explicit enough . . .

I had assumed that this pattern of behaviour would alert my dear mama that something atypical was going on, worthy of investigation. Instead, she talked a lot but studiously said nothing.

Fortunately, even her intense fortitude could only last so long. After three days of auditory assault by Mr Garfunkel, she eventually gave in to her burning curiosity in the only way she knew how. She eavesdropped on one of my daily telephone conversations with Francis.

There were two phones – one in the lounge and one in my parents' bedroom. As usual, I was taking advantage of the relative privacy of the bedroom. My mother must have been sitting on the long sofa in the lounge, aching to pick up the other extension. This she evidently did, but *extremely* quietly. This didn't prevent the line from suddenly sounding slightly more echoey, a symptom that I recognized immediately. I carried on with my lovey-dovey conversation regardless. This seemed as good a tactic as any. The line remained a little bathroomy for about ten minutes, then, as professionally as before, my mother replaced the receiver and the sound reverted to a quality that the GPO who'd installed it would have been proud of. Half an hour later I finally rang off. And wandered down the hall to confront my mother.

Who said absolutely nothing.

For almost twenty-four hours. And then, out of the blue, in the kitchen, she broke.

'So, does all this mean that you will be adopting the descriptor "*gay*"?!'

To the benefit of innumerable children coming out to their caring parents, it is fortunately surprisingly difficult for someone to spit out the word 'gay' – their lips and upper palate are in the wrong place – but my mother made a valiant effort nonetheless. With that said, the exertion had clearly exhausted her, and she stood in pale silence, her lips very slightly parted and curled inwards, for a good five minutes as I calmly explained. This was a moment I'd been waiting a long time for, prepared for. It was important she genuinely understood and so I tried to be at my persuasive best.

'I see!' was her eventual concession.

I wasn't yet sure that she really did, so I continued for another ten minutes, filling her in on what my teenage life had really been like and what I hoped my adult life would be like going forward.

'I see!' This time it was far more efficient-sounding. It was the tone she always used when she finally understood something. 'Well, you are clearly horribly confused.'

I didn't feel upset so much as frustrated.

'No! I'm *really* not.'

'Well, you obviously are!' This with a patronizing fake half-laugh. 'You can't possibly know what you want at your age.'

'Really? Didn't you know that you fancied boys when you were my age?'

She seemed slightly flustered by this. But only for a second or two.

'No! I wasn't even thinking about things like that back then.'

This seemed rather unlikely given that, by my calculation, at around my age she'd been about to marry my father. But I let it pass.

'Well, I was certain at thirteen.'

'Rubbish! You've been led astray by this revolting man. He disgusts me. He's an older man, of course?'

Her uncertainty seemed disingenuous given that she'd probably studied Francis's photo through a magnifying glass.

'He's two years, six months and fourteen days older!'

She tossed her head and grunted at this new insight, as if age was suddenly irrelevant to her argument.

'That's beside the point. And you say he works in a *hotel*?' Once again, she found herself hampered in spitting out a word, this time one beginning with 'h'; once again, she gave it her best shot.

'He's wonderful at it. And it's a lovely hotel.'

'You said it was *exclusively full of homosexuals*!'

'That's what it's for!' I was growing exasperated. 'It's a gay hotel. It's a sanctuary for people *like me*, where we can escape from the ignorant, hateful bigotry that even *you* are displaying right now!'

She flinched as if I had slapped her in the face. Her nostrils pulled in as she drew a deep breath. Then, in a slow controlled voice:

'Your behaviour *repels* me.'

A week before, that blow would have devastated me. Instead, armoured by love, it bounced off and I barely felt it. Mentally, I shrugged. It was a lost cause, at least for now. My riposte to her attack sounded even calmer than she had:

'And your behaviour today utterly repels me.'

With that, I turned away and walked to my room.

For the next three days, like a couple of identical magnetic poles, we dutifully repelled each other. We never spoke. When we saw each other, we veered apart. My father and I *did* occasionally speak – but only, of course, about inconsequentialities.

'Peter!' I'd seen her in the dining room and carried on walking past the door in the direction of my bedroom. Although, in normal life, she never called me by my name, this was at least an acknowledgment of my existence. It seemed churlish not to respond. I retraced my steps and stood in the doorway.

'Peter – I've decided I would rather try to understand you than grow into an embittered old woman.' This was clearly a pre-prepared speech. But at least it was speech. 'You have your father to thank for this. He said: "He's still our Peter." And, of course, you are. That's why all this is such a shock.

Neither Da nor I want you to feel isolated and unsupported like you've felt as a teenager. We want to help.'

I felt a sudden surge of relief; I hadn't realized I was so desperate for her approval. Ma's tone of voice softened into a soothing maternal caress.

'But to do that, *you* need to help us. You've got to stop all this silliness. Whatever your . . . predilections . . . we will try to help you get through them. But you *cannot* risk flaunting them. If anyone gets the slightest hint of anything untoward, then not only will the whole family turn against you, the whole *world* will turn against you. And Da and I won't be able to protect you. So you *must* meet us halfway. You simply *must* give up this "Francis" and carry on as normal. It's the only possible way.'

'What?'

'It's the only choice.'

'I'm *never* going to give him up!'

'He's a childish infatuation! You've only known him a few days, yet for someone you've only just met you're about to destroy your whole life, *everything* you've worked so hard for, *everything* you take for granted. It's a cruel, cruel world out there. You do *not* want it to turn on you. You would be ruined. And Da and I would just have to watch and let it happen.'

I remember feeling shocked – as I was no doubt meant to. For some reason I'd always assumed that my mother would, as ever, be loving and supportive when she at last learned the truth about me, and it would be my father who struggled to accept it. In practice, they were *both* now enacting the nuclear option. And taking for granted that I would capitulate.

I remember wondering for a fleeting instant how her father would have reacted if he'd still been alive. He'd been my favourite grandfather. And he'd been an atheist. He'd had a harsh Victorian upbringing thanks to his coal-miner father, who subjected him to beatings. But he was so bright and so determined that eventually he broke free, got a degree in geology and won the hand of my then-beautiful grandmother, who had the strength of character to marry 'very far below her' simply because she loved him. Now, his freethinking influence was long gone. I chose my words with care.

'Let me get this straight.' An unfortunate turn of phrase given the circumstances, but I persevered. 'According to you, either I give up Francis or I give up the family, everything I'm familiar with and all my prospects?'

'Don't you see, darling? I know it may feel a little harsh, but you have absolutely no alternative. You have to give him up. It's the only possible sensible thing to do.'

She was right, of course. It was a no-brainer.

'A future with Francis *or* everything from my past?' I pretended to calculate the trade-off. 'I choose the future!'

She didn't attempt to change my mind or even express surprise. In an instant, her face locked into dismissive resignation and disappointment.

'You're as stubborn as your grandfather was!' For the first time since our battle of wills had begun, she broke eye contact. Then she looked back at me to fire her final fusillade: 'You are giving up *everything*!'

I actually managed to smile as her volley failed to strike home.

'He's worth it!'

In contrast with trying to persuade my parents to support me, persuading Imperial College to put my computing science degree on hold was easy. I convinced them that – before selecting my specialization for my final year – I needed to take an indefinite sabbatical to explore my preferred career path and confirm it was what I really wanted. My only challenge was to come up with a suitable career path that was simultaneously believable and alluring, but also obscure enough that the only way I could learn about it would be to self-tutor, anywhere in the country, even Devon.

The second most important decision of my life (as arbitrary and unlikely as the third most important decision of my life) was largely made courtesy of Dr Isaac Asimov. On the train down to Devon, and then back to London, I'd been reading his sci-fi book *I, Robot*. I loved it. More importantly, robots involved lots of computing, they were a believably alluring career path and they were *so* abstruse that even Imperial didn't yet have a robotics department. Perfect. Within two weeks I had negotiated an open-ended sabbatical 'to study robotics'. It was the day before my twenty-first birthday.

That night, as Big Ben struck midnight, I was standing on Westminster Bridge with a large suitcase and a group of my college friends. We popped the cork of a bottle of champagne and drank to my adulthood and the fact that for the first time it was now at last legal for me to fornicate with another man. Once that bottle and the next were empty, we all made our way to the Tube and there hugged our goodbyes. I waited for the late-night service, made my way to Paddington, caught the milk train, changed at Newton Abbot and eventually arrived at Torquay, and Francis welcomed me to my new life.

Devon in the late spring and summer is perfect. At least, it was in 1979. It was an idyll. Francis and I were experiencing the euphoria of early deep love, and I laid down more happy memories then than in any other four months of my life. I adored the sea, the cliffs, the beaches. I adored Dartmoor and the fact that in those days we could lay a blanket on a remote sunny hilltop and remain undisturbed other than by a grazing wild pony. I adored the exotic bohemia of the Cliff House clientele and my first ever exposure to a comprehensive cross-section of society. I adored the fact that when (over a luxurious cream tea in what would later be renamed 'Bovey Castle') I explained to Francis my views about humanity increasingly turning into cyborgs, and that it would happen within our lifetime, that it was science not science fiction, his immediate reaction was delight that in that case we would be together forever. Above all, I utterly adored Francis. So when he suggested I finally meet his mother, I very much wanted to make a good impression.

'Eeer merbuck uzzelbee gwain kwop drektlee!'

She was smiling at me encouragingly, so I gave her a huge smile back.

'Great!'

As she bustled into the kitchen, cursed the Jack Russell intent on tripping her up and began searching for her door key, I appealed to Francis for a translation.

'Oh, you'll pick it up soon enough. It's a sign that she likes you that she's speaking Devonshire. Otherwise she'd be "speaking posh". All she said was: "Here, my buck" (that's "Hello, young man"), "we will be going to the Co-op directly" (in other words, "soon").'

'What's the Co-op?'

'The Co-operative Society.' I continued to look bemused. He helpfully added: 'It's a type of shop.'

By August, every day was unending sunshine, the Cliff House garden was in full bloom and academia felt a universe away. I had indeed studied robotics and been passionate about it. And, under different circumstances, I would have chosen to specialize in it. But not now. Not in this life.

Instead, I had decided to give up all thoughts of completing my degree, let alone going on to get a PhD. I wanted only to be with Francis. And he *loved* his world centred around the Cliff House. So did I. We were together, forever, in Devon. That was all that mattered. Then, one night, he shattered my dream as we were both sitting up in bed.

'I'm really sorry, but I can't stand it any more. You've got to leave and go back to London.'

Contra Mundum

For a moment, my brain went into shock and didn't properly process the next few sentences as he continued his rehearsed monologue.

'I've never known anyone who could do a PhD. I never understood what it meant. I never appreciated what an amazing opportunity it was. But now I do. And now I know you. And you would be brilliant. It's what you should do. It's what you're designed to do. And I love you far too much to get in your way. You've *got* to go back and finish your degree and then get your PhD!'

I began to protest, but he lovingly put a finger on my lips and carried on.

'We'll be poor for five or six years till you finish, and I don't know where we'll live' – We? We'll be poor? We'll live? – 'but we'll find a way. I'll find a job somewhere, doing something. And we'll be together every step of the way.'

Yes, of course 'We' – there was never any doubt. My brain finally reconnected with my mouth.

'But, darling, how can you possibly leave here? It means everything to you. It's the only job you've ever loved.'

'And I love you infinitely more. You were willing to give everything up for me; now it's my turn.' He paused and for the first time looked concerned. 'You do *want* to be with me forever, don't you?'

I couldn't quite mask the residual incredulity in my voice, even as I tried to turn my innermost fear into a joke:

'I thought you'd never ask . . .'

A month later, as the late summer began to cool, we began a rotation around an unlucky selection of our friends who happened to live in London – a few nights in the spare room of one, a night on a couple of sofas at another, and so on. It went without saying that none of the capacious spare bedrooms lying idle in the oversized properties belonging to my extended family were potentially available, let alone my old bedroom in Wimbledon; but, after a particularly uncomfortable night, I'd double-checked. I shouldn't have wasted the money in the phone box.

Every morning, we'd get up early, run to the nearest newsagent, buy a London paper full of rental ads, pore over them all through breakfast, pick up our small plastic bag of coins, head to an unoccupied phone box, squeeze inside and work through all the ads we'd circled. Most times, it was a very simple conversation:

'Sorry, it's already gone.'

Occasionally, we got an appointment. Only to find that the accommodation wasn't what we'd been told, or where we'd been told, or for the amount we'd been told, or simply that the landlord didn't want to rent to two men.

Eventually, at just about the time we had overextended our welcome with every eligible friend and acquaintance, we at last found a place to rent that we could afford and from which I could walk a couple of miles to get to Imperial each day and so save the cost of a Tube ticket. It was a genuine garret flat – up in the roof space of a five-storey Victorian townhouse, with small windows overlooking the London chimney pots. It was gloriously romantic.

It was also surprisingly cold, we discovered, as the autumn nights got chillier. By the time we found ourselves starting to actually shiver in the evenings, I'd resumed my undergraduate degree and Francis had found a job looking after adults with extreme learning difficulties. Although Francis had grown up feeling cold indoors, it was a completely new experience for me. I rapidly concluded it was an experience I could do without. We bought a large roll of Sellotape and proceeded to tape up all around the windows to seal the gaps through which the increasingly wintry drafts were stabbing.

Now hermetically sealed into our ergonomically compact all-purpose living room/dining room/kitchen/lounge/scullery/larder, we would huddle around the small gas fire, imagining how warm it would feel when we finally lit it. We had rationed ourselves to one shilling a day to feed the gas meter – twenty days for one pound.

The only problem was that, having cooked dinner on the gas hob by the sofa, even if we kept the gas fire low, we only got a couple of hours' heat. Of course, by then we typically had headaches from having used up much of the oxygen in the room, so heading to a freezing bed while we were still warm enough to slowly heat it up seemed like a sensible way to end each short evening.

But, while the gas fire was hissing out warmth, and our fingers slowly thawed, we would talk. Almost every evening, we planned and schemed and dreamed of the future. Once I'd got my PhD – maybe as soon as five years' time – we would be free to do X. Or maybe Y. Or maybe both. As well as Z. Whatever it was, it was going to be awesome.

Occasionally, of necessity, we talked about the trials and

tribulations of the present. We were, we realized, now totally on our own. Just as my mother had promised, my extended family had efficiently and surgically cut themselves off from me in order to avoid any possible taint of infection from the rainbow sheep of the family.

Part of me, I realized, had assumed that at least some of them would not ostracize me quite so vehemently and unrelentingly. But I'd been wrong. Hurtful though this was, I knew it was worse for my parents. They had to suffer the ignominy of being pitied.

'It's their loss,' Francis concluded.

'The person they all said they loved obviously wasn't me. So I haven't really lost anything. And at least Ma and Da were civil.'

'They were very nice! Considering . . .'

After some time, diplomatic relations had been opened with a cautious invitation for Francis and me to come to Wimbledon for tea. It had all been very civilized, and Ma had used an excellent china tea service, a Georgian silver teapot and (I couldn't tell if she appreciated the irony) served fairy cakes. Beneath my parents' façade of extreme politeness was a yawning chasm of discomfort. But their sense of duty meant they were at least trying. So there was hope.

'It's us against the world,' I summarized. *'Franciscus Petrusque contra mundum!'* It was the first part of our mantra: Francis and Peter Against the World.

'Then the world had better watch out!'

'Coniuncti vincemus!' I chanted, completing our call to arms: Together We Conquer.

One of the huge upsides to our accommodation was that although we lived in relative poverty we also lived in the centre of London. We could walk anywhere. And so it was that approaching midnight on New Year's Eve we found ourselves walking down Regent's Street towards the crowd congregating in Trafalgar Square for the celebrations of not just a new year but a new decade. We were feeling jubilant. And we were walking hand in hand. We'd never seen two men walking hand in hand; it seemed a suitably radical way to welcome the eighties.

We waited in the throng for the stroke of midnight. According to my watch, which I'd carefully synchronized with the pips on the radio before the six o'clock news, we were less than a minute away. A party of revellers close by prematurely started cheering. But I knew they'd got it wrong. I still had time.

It felt important to put into words what had been slowly dawning on me over the preceding nine or so months. When Francis and I had met, I'd been fighting alone for so long that – if only to keep up my morale – I'd kidded myself that I was fully self-reliant. Ever since I'd rebelled at sixteen, I'd prided myself that I could take on anyone. Anything. On my own.

Gradually, as Francis had begun to polish some of the rougher edges in my character, fill in some of the more gaping holes in my general knowledge, I'd realized how restricted my exposure to the real world had been. Recognizing just how weak and ineffective and narrow in outlook I actually was, despite all my inner resilience and self-belief, had been an insight I'd struggled with for at least six months. It was the shedding of a layer of protection, of sorts. Once the revelation had finally sunk in, however, the implication for the rest of my life was obvious. It was by far the least scientific, the most human, the most important decision I ever made.

It was much more than just choosing to be with Francis forever, recognizing him as my soulmate, knowing that any sacrifice to keep him was worth it. It was more even than letting down the formidable defences I'd built since early childhood, willingly becoming vulnerable, being open to change. At its passionate heart, my life-defining decision was about enthusiastically relinquishing parts of my personality so as to *meld* with Francis, as an equal partner, as part of me, as part of him. To become greater than two. To become one.

I tried to sum up the logic to Francis in the final seconds of the year in which we had met:

'Always remember, I'm nothing alone. But whatever the universe throws at us, together we are *invincible!*'

Dartmoor

It started with a tickle in my throat, suddenly, with no warning. The sort of tickle you don't even think about. You just cough and that gets rid of it. We'd already passed Haytor Rocks and were now cruising down a slow incline past the occasional grazing pony, with staggeringly beautiful Dartmoor stretching to the cloudless horizon. A brilliant blue sky sliced across a verdant moor spiced with shocks of yellow and purple courtesy of bracken, gorse, grass and heather. I coughed.

The tickle was still there. Was that five cows in the distance, or five ponies? I coughed. Ponies – I could see their necks. I coughed. Oh, there was a new-born foal! I coughed. I've got a tickle, a bit of phlegm or something. I cleared my throat, using the time-honoured method of harrumphing like a respectful valet trying to attract his master's attention. The tickle was still there.

After less than a minute, I was coughing uncontrollably, every few seconds.

'Are you all right?' Francis didn't sound worried so much as intrigued.

'I ca–' Trying to speak turned out to be a catastrophic mistake. Up until that moment, in the seconds after each cough I'd been able to breathe in enough to cough again. I'd been in balance. But now I'd used the precious time between coughs to lose still more air by starting to speak. And then, before I could draw in any breath at all, I coughed. Always a double cough. Beyond my control. But my lungs were already empty. My instincts took over and I involuntarily gulped to try to get enough oxygen. And failed.

My airway was partly blocked. It was as if I was being strangled. I whooped. It's the sound they use in movies when someone who's been underwater far too long finally breaks surface and triumphantly draws a life-giving breath. In medical jargon, it's called 'stridor'. Hollywood tends only to use that painfully strained sound once, to show how near-death the situation was. Then the hero breathes normally. I didn't. Instead, I coughed. Worse, I coughed before I had noisily drawn enough breath even to cough. Desperately I whooped again. Coughed too soon. Whoop. Cough-cough. Whoop. Cough-cough. Whoop. My DNA told me to panic.

Francis veered over to the verge and emergency-braked. Cough-cough. Whoop. My stridor was now far worse – an urgent yelp of anguish. Cough-cough. Francis shot his right hand out, strained over and scrabbled in the glove box for a plastic bottle of water. Whoop. He twisted the cap off. Cough-cough. Turned to me and grabbed my hair with his left hand. Whoop. Pulled my head back. Cough-cough. Waited for the immediate whoop to finish, forced the bottle-neck against my mouth and, as I weakly coughed (I had no breath left to do other than croak), he squeezed some water into my mouth. Most of it spilled on to my lap and immediately made me feel like I'd wet myself.

But I swallowed a very small amount. And for the first time in the last two minutes drew a slightly bigger breath than the one before. I still whooped, but my lungs didn't feel quite as painful, the fire inside my chest didn't burn quite as fiercely. Francis forced more water into my mouth. I was vaguely aware of him murmuring words of calming reassurance. My uncontrollable cough became instantly controllable as soon as I could swallow a proper slug of water; the stridor took a bit longer, gradually evolving from a whoop to a painful gulp to a gulp to an urgent breath to a breath.

For the first time since my diagnosis, I felt in shock. This was clinical shock. I felt cold, and clammy, and scared. What if Francis hadn't been next to me? What if we'd been home and he'd been downstairs and I hadn't been able to call out? What if we hadn't had some water in the car?

'What was all that about?' Caring. Compassionate. Worried.

'I couldn't breathe!'

'I guessed that much! But what caused it?'

A good question. A good *scientific* question. As I pondered my response, I felt myself calming down.

'You know that flap at the back of your throat that closes over your windpipe to stop food going down the wrong way, the epiglottis? Well, I felt something on it, I don't know, a bit of phlegm, or a crumb of the sandwich we just ate that got dislodged from a tooth, or something, whatever; anyway, I felt it tickling away and I couldn't stop it making me cough, and I couldn't clear it and I couldn't breathe.'

'Well, we know that's the direction of travel.' Matter-of-fact. Realistic.

'I know. I just didn't think it would be so soon. Or like that. Dysphagia normally starts with problems swallowing or being unable to control your gag reflex.'

'Well, at least you've always been very good at controlling your gag reflex. Should we turn around and do this tomorrow? You're looking a bit pale.'

'Would you mind?'

He immediately pressed the ignition. 'Of *course* not!'

As Francis executed a U-turn, gently persuading an errant sheep on the other side of the road to move over a bit, I reflected that I was using the words 'sorry' and 'thank you' a lot more these days. We headed back over Dartmoor, with glimpses of sea in the far distance and me left utterly confused about how *anyone* could be drawn to autoerotic asphyxiation . . .

'Let's try that again!'

It was the day after my epiglottis episode. Part way down the steep hill, Francis carefully pulled off the road on to a relatively flat area of scrub, avoided a particularly perilous-looking rock and parked looking down into the beautiful valley, ponies grazing close by, cows grazing over a mile away on the hills in the distance. Below us was the village of Widecombe-in-the-Moor, dominated by the towering spire of Saint Pancras Church, incongruously large yet perfectly beautiful. Francis turned off the engine, opened all the windows fully and we sat for a while as we listened. A very distant tractor. The occasional moo. Bleating. The fleeting buzz of a bee as it flew past and then was gone. Peace.

'So, apart from your robot voice –'

'It'll actually be a *cyborg* voice.'

'OK, apart from your cyborg voice and your cyborg avatar and being in virtual reality and using AI to control everything, what exactly are your plans to change the world?'

At least by now I'd had a bit more time to get my thoughts in order.

'Well, I see the things we've already talked about as three distinct research streams: verbal spontaneity, personality retention and – for want of a better term – virtual liberation. I think there are four more –'

'*Four?*'

'They're very simple. But they're importantly different from each other. Look, the next one I call robotic mobility, which is pretty obvious.'

'You'll be a robot.'

'Well, no, I mean CHARLIE will be robotic.'

Although it wouldn't be delivered for several months, I'd already named my recently ordered Permobil F5 Corpus VS 'CHARLIE' – short for cyborg harness and robotic life-improving exoskeleton. It was based on a standing electric wheelchair, but I'd persuaded Permobil to help turn it into something amazing. I wanted it one day to be the non-biological core of Peter 2.0.

'The point is, my hands will eventually stop working; I can already feel my little finger going. So, realistically, within a year or so I'll no longer be able to control CHARLIE's every action with a joystick. I need him to take on far greater responsibility. But, exactly as with the other three research streams, the very low speed of my eye-tracking input system is my weakest link so I need CHARLIE to demonstrate a huge amount of autonomy in getting me from point to point – just as my legs and whole body used to.'

'You said that the other day – about delegation, just like your brain does when you're walking.'

'Yes, exactly. I was talking about using AI to offer options for how to move my avatar in virtual reality. But you're completely right. It's the same thing in "real" reality. They both need loads of AI to work properly. Robotic mobility is just an AI challenge on wheels.'

'Are you practising sound bites?'

'Just in case.'

'Well, you never know . . .' Always supportive.

'For example, I'm thinking that when we're outside the house onboard our WAV –'

'Remind me. WAV?'

'Our van. Now it's been converted, technically it's become a "wheelchair-accessible vehicle". So, there I am in the WAV on the drive, and I use my eyes to click on an icon for 'bedroom' and that's it – everything else is automatic until I'm safely by my bed having got off the WAV, gone through the front door, up in the lift and through into our bedroom.'

Nobody had ever explained to us just how expensive MND was. With inestimable timing, one year before my diagnosis Francis and I had bought a house next door to nephew Andrew – David's brother, now long out of the army and handling VIPs flying into Exeter Airport on private jets. Also living next door were Andrew's partner, Laura, our soon-to-be-three great-nephew, Ollie, and a soon-to-be-zero great-nephew or -niece.

Our living arrangements were a source of joy. Except for one thing: our house was on three levels and even our garden was on two. As my walking had got worse and, strictly in the privacy of my own home, I'd reluctantly resorted to using a walking frame, Francis had worked out a way of installing a lift large enough to take a full-sized electric wheelchair and a carer. It was a major undertaking, lost us two of our bedrooms, took three months and cost £30,000.

Then there was the WAV. We'd never thought that once I could no longer transfer from a passenger seat to a wheelchair we'd need a vehicle that could take the wheelchair. Only then, when we'd looked around, had we discovered that almost every such vehicle looked like either a box on wheels or a hearse. Eventually, we chose a glorified people carrier. It was hardly going to take us off-road on to the moor – and certainly not back again – but at least it had a modicum of power and didn't rattle too much at speed. But it had to be converted, a side-lift added. The complex conversion alone was another £30,000. Our 'rainy day fund' was empty. And the deluge continued.

'That's really important! You being able to get around the

house easily on your own, and in and out of the WAV, would be brilliant. But what about when you're outside – somewhere you've never been before?'

'Even in unfamiliar territory, I want to be able to travel fast but safely using a sophisticated collision-avoidance system, just like they're devising for cars. I'm imagining being able to speed through an obstacle course or safely make my way through a showroom of porcelain vases –'

'You are *not* going to speed through the Dartington Crystal showroom!'

'It's a thought experiment! But now think of this: I'm speeding through the obstacle course or china shop (but not Dartington) and I've got a VR visor obscuring my eyes. And what I'm seeing is augmented reality – my intelligence amplifier at full pelt.'

'Wait, wait, let me get my head around that. You're in VR, but what you're experiencing is the real world?'

'Yes, because to me they'll feel remarkably similar. After all, it's the AI that's doing all the difficult stuff – I'm just indicating what final destination I want.'

'Got it! So, it doesn't matter whether you're looking at the real world or at a VR visor because you're not steering anyway?'

'Precisely so. And imagine, on another occasion, I'm using the same system to teleport into a real-word meeting (not a VR meeting on another planet) and I'm occupying a remote telepresence robot.'

'You're losing me . . .'

'You know, a remote-control robot. I'd see and hear what I would if I were in the room, and everyone there sees and hears my avatar. And I can steer around and follow people down a corridor just like little Ollie does with his remote-control car. On another occasion, maybe I've teleported into a drone flying across Torquay Harbour, looking down at you and me from my extra-terrestrial body.'

'I wouldn't say things like "teleport" or "extra-terrestrial body" to people – it'll scare them off. Why don't you just say "remote control". Keep it simple.'

'I think, or at least I hope, some people will *love* the idea. The point is that what I'm talking about is far more profound than just remote control. Imagine, for example, that CHARLIE is making his way through the house as before, but I'm in bed – experiencing through VR what I would if I were sitting in CHARLIE. Do you see? Do this right and it'll totally start stretching our perception of what "reality" really is . . .'

'Well, thank goodness your brain is suited for this sort of thing. I'd hate it for myself! You know that I wouldn't be able to cope with the technology.'

'That's one of the many reasons I'm glad I've got MND not you!'

'Not least because you couldn't cook us a nice dinner!'

My brain was suddenly racing with a surge of ideas it proposed sharing with Francis, but I didn't want to redirect the conversation. Too much.

'You do know the technology will get easier and easier for you to use? In another decade, it'll be really easy and really powerful.'

'I know. But I'm not you. I'm perfectly happy to scandalize society by growing old disgracefully – married to an ageless robot.'

'Cyborg. But just to throw in a thought: what if all this works?'

'How do you mean?'

'I don't want to eventually get more and more freedom and watch you getting less and less. If I get to fly across the harbour, I want to fly with you!'

'You know I don't like heights.'

'Metaphorically.'

We pondered this deep thought for a few minutes, enjoying the tractoring, mooing, bleating and buzzing until Francis pressed the ignition and we edged back on to the road, with me wondering if this might be the very last time we went off-road.

Five minutes later, we passed the Old Inn and made our way out of Widecombe down the long single-lane road with passing places that would eventually lead us to the crossroad we needed for Hound Tor.

'So, what's your next set of wild ideas?'

Wild Ideas

'I think I'll need to have a partial exoskeleton . . .'

'There you go again! You'll scare people saying that. They'll think it's all science fiction. Or they'll find it intimidating.'

'But it's *not* science fiction, or at least it doesn't have to be. Without some way of moving my limbs, they'll never move. And worse, I'll never feel anything again. Never touch anything. So, I'll need some form of exoskeleton arms and gauntlets with gaps for my fingertips and palms so I can touch surfaces and distinguish textures. Because otherwise it's insane that I'll keep all my sense of touch and never use it. I want to be able to touch things. I want to be able to reach out and touch *you*!'

'So long as you don't accidentally knock me out!'

'You know, already people don't touch me any more.'

'*I* touch you!'

'Yes, of course *you* do. But I've noticed most people don't. They'll shake hands with you, even give you a hug, and then they'll turn to me and give a funny wave.'

'I don't think they quite know what to do. Whether it's right to invade your space. No one teaches us at school how to touch people in wheelchairs. Especially when they are unable to move . . .' he gave a slight pause but his blue eyes didn't flicker, '. . . like you will be.'

'Well, there, that's exactly my point, that's one of the reasons I need to be able to move, to hold out my arm to send the signal that I want to shake hands. And, by the way, I'll also need an exoskeleton neck that's capable of turning and nodding my head so I can at least look around. I mean, how frustrating would it be to only ever look forward?'

'You mean like everyone else with MND?'

'And tetraplegics, yes. But – there's a theme here – for even this partial exoskeleton to work, it needs sophisticated AI to allow me to control it through instructions using only my eyes. So, as ever, I'm visualizing being offered options – exactly as when I'm in immersive VR. In fact – what am I saying? – it really will be *exactly* as when I'm in cyberspace because they will be the *same* options. Think about it: why should it feel *any* different in the physical world?'

'Because the physical world is real!'

'Yes, I know, but whether I'm turning a handle to open the door, or coordinating movement between both my arms or reaching out and touching someone I love, I will do them all in exactly the same way in cyber-reality as in space–time reality. Do you see? *Both* will become my expanded reality.'

'Jay's Grave!'

Francis braked to a crawl. I'd almost missed it, on the driver's side, at a crossroad chosen long ago so that Kitty Jay's ghost wouldn't know which way to go to reach the village. Her ancient grave – necessarily on unsanctified land as befitted someone who'd hanged herself – still had fresh yellow flowers on it, just as it had every time we'd checked ever since that first week I'd moved down to live with Francis. Just as, so the legend went, it had for centuries, 'placed by an unknown hand'.

'Yellow flowers,' Francis noted.

'Yellow flowers,' I confirmed.

Francis accelerated away. 'I can certainly see why it's important for you to touch things with your own fingertips.'

'Well, this is where it actually starts to get rather interesting.'

'At last!' But he couldn't keep a straight face as he'd planned and started laughing. I smiled back, although his eyes were on the road.

'Most people assume that an exoskeleton's only purpose is to interact with the physical world. But in my case, it'll serve an equally important second role – and this is really clever –'

'Though you say so yourself.'

'Because it'll allow me to *feel* interactions in the virtual world. Because none of my voluntary muscles will work, *my* sensations (when my exoskeleton moves my arms, experiences resistance, hits something, is weighed down, bounces back) will be identical whichever reality I'm in. So, I may look like I'm just wearing an innocuous exoskeleton but in truth I'll be wearing the ultimate cyber-suit that will be the envy of any futuristic VR gamer!'

'And what am I going to do while you're spending all your time in virtual reality?'

'You'll be in there with me!'

'I will *not*! I like being in the real world. I don't *want* to be in cyberspace.'

'Brad Pitt'll be there . . .'

'Brad Pitt, you say? In his twenties, or now?'

'Whichever your heart desires.'

'Both, please. At the same time. It's an excellent idea! Sign me up!'

'Well, that's good because the next stream of research builds on all that; it's an all-access pass to cyberspace.'

'You'll lose people if you call it that!'

'Yes, but that's what it is. In essence, I'm imagining seamless access to anything electronic, with some form of intuitive navigation across the *whole* of cyberspace.'

'You mean, the whole of the internet?'

'*And* lots of the devices and systems that in the near future are going to be *connected* to the internet.'

'Is that the "internet of things" that you've been banging on about?'

'Exactly! And I want to become one of those things! Think about it: by tying into every possible sensory device on CHARLIE – stuff that can let me see, hear, feel, even smell things – I'm hoping that over time my boundaryless immersion into the cyberworld will feel like having a massively extended body.'

'Is that a form of cyber-obesity?'

I laughed. 'What I mean is that eventually, if I do it right, parts of the internet and things connected to it may end up acting like substitutes for my original paralysed body. Instead of interfacing with my environment, increasingly I will *become* my environment.'

'Didn't you write something along those lines in *The Robotics Revolution?*'

'Yes, I've always believed it. It's obvious. Thanks to the plasticity of my brain, over time, cyberspace (and any of the physical world accessible through cyberspace) will feel like *me*; sending an email or calling an elevator will be just like raising a finger or raising an eyebrow used to be.'

'And you'll no longer feel trapped in a paralysed body!'

'It's not just that I'll no longer feel trapped in a paralysed body, I'll no longer *be* a paralysed body. My new body will potentially extend everywhere – across not just the physical universe but also an infinite range of possible *virtual* universes as well. Can you imagine?'

'I struggle with programming the toaster! But yes, you'd be in your element.'

'And I'd be breaking the barriers between machine

intelligence and human intelligence. Instead of competing, we'd be melding.'

'Like you always said ... But what about recording things on the TV? I'd feel a lot more comfortable about all this if you'll be able to stay in charge of our evening entertainment.'

'Of course, in my immediate environment (especially if it's very familiar, like home), I'll have intuitive eye-driven control of everything I used to accomplish with my hands – definitely controlling the TV, but also household appliances.'

'You don't use those *now*!'

'Yes, but in principle. And calling the lift and opening doors. And naturally, because I will by this stage have full access to machine translation, there's absolutely no reason why any of my conversations can't be in Chinese.'

'You've *never* spoken Chinese!'

'Exactly! Despite the fact that I don't know Chinese, as Peter 2.0 I'll be able to use my avatar to hold a video Skype interview in Chinese *and* Japanese – at the same time!'

Francis slowed at a junction, turned left and pulled into the long parking area in front of Hound Tor, famous as one of the rocky locations referenced in *The Hound of the Baskervilles*. Before, Francis and I had always got out of our car at this point, and we'd walked across the rough tussocky ground to the Tor. Sometimes we'd climbed the granite outcrop, tricky in places if it was at all wet, until we could stand at the very top, viewing the immensity of Dartmoor in every direction. Once, on a summer long ago, at a time before there were so many tourists, we'd even spent a passionate ten minutes on the summit.

This time, we just sat. And Francis opened the windows. And we watched an old couple getting out of their car and slowly walking across to the rocks beyond, one helping the other. I felt like turning to Francis and saying, 'I'm so very sorry.' But I knew he knew.

'You see how even that geriatric pair can make it across to our rocks, but I can't?'

'Now, that's not how we talk! We've got to focus on the future, not the past, on the positive –'

'I know, I know, that's not what I meant. My point is that those two symbolize the need for my final research stream – what I call my all-access pass to the physical world.'

'Oh, I love the sound of that. I so want us to go up the Nile again one day. And we never did China and the Far East together. Or Sydney Harbour. And we haven't been able to return to the States for ages. And –'

'I know, I know, there's a lot of the world we still haven't seen, or that we want to see again, and that's my point. That's what this last bit is all about. But the ideas get a bit complicated, so bear with me.'

'What's complicated about the idea of getting wheelchairs to cross moorland or go up stairs or get onboard planes?'

'Well, yes, that bit's simple. Well, actually, it's really hard to do. But conceptually, it's simple. And I really do think there's a massive amount we need to tackle head-on to obliterate the huge barriers there currently are to disabled access.'

'It's insane that the main reason we won't be able to take a plane anywhere is that no airline will allow CHARLIE in the cabin.'

'And his batteries would have to be disconnected if he was in the hold. And even if I could survive separation from my physical and mental life-support system through a long-haul flight, based on current statistics the chances are that CHARLIE would be damaged by the time he was returned to me.'

'I've just thought: as a cyborg, will you have to be put into flight mode?'

'I never thought of that! I wouldn't be able to speak. I couldn't communicate if something was wrong. And it's sloppy science anyway for the airlines to claim that mobile phones or laptops are dangerous to have on – it's just legally convenient. Gagging me like that would be an infringement of my human rights! It's discriminating against me because I'm a cyborg –'

'We are *not* going to start campaigning for cyborg rights!'

'I think we're going to have to! There's going to be the same sort of casual, institutionalized and often unintended discrimination as there was against gays in the seventies and eighties –'

'And the nineties, and the noughties . . .'

'Exactly! The law tends to lag way behind technology.'

'Brilliant!' He did *not* sound enthused. The more I thought about it, neither was I. But needs must.

'Back to the plot: I'd love to be able to easily navigate a typically hostile urban environment, accompany Andrew and Co on a country walk, safely climb stairs, board a boat, traverse icy or snowy surfaces –'

'That's something to focus on.'

'But – and here's where I know you'll instinctively think I'm going too far, but I'm *really* not and it's *really* important – why should I straitjacket us by specifying that my biological brain always needs to risk coming along for the ride?'

'There you go!'

'No, listen, it's crucial. Here's what I mean: I'm predicting that, eventually, whether I am in physical reality, augmented reality or fully virtual reality will feel as irrelevant as whether a movie you and I are watching is in a cinema, live on TV or downloaded to a laptop.'

'They're all very different . . .'

'Yes, but at the end of the day all we really care about is the experience of the movie. In the same way, whether I'm actually in physical reality or virtual reality, all I care about is the experience of reality. I'm perfectly happy that it's ambiguous reality.'

'As you've always said, we can't prove that we're not all living inside AI already, so I suppose if you can't really tell the difference then it doesn't really matter which reality you're in.'

'That's exactly it! But that has hugely liberating implications! Think about it: in my ambiguous reality, you and I can once again be hand in hand on a mountain peak, and I'll feel a gentle summer breeze and the early-morning sun on my face, but maybe I'm safely at our home in Torquay and you're beside my backup CHARLIE, a doppelganger that electronically relays me to the mountain peak and the mountain peak to me.'

'You mean, I'm actually there but you're not?'

'Maybe. Or maybe my physical body is there, but on a four-legged walking machine. Or maybe the walking machine is another form of doppelganger and I'm in Torquay. Or maybe you're using an ultra-portable relay that's primarily a drone that follows you at eye level and lets me feel as if I'm by your side. Or maybe we're both in Torquay . . .'

Silence, as he processed all this. In the distance, the old couple had made it to the Tor and were triumphantly sitting on a rock, surveying all around them.

'Is all that a very long way off?'

'Not necessarily – it could be around the corner. It's all technically feasible today. It's just that no one's doing it. I want to nudge things in that direction. I need to find a way to get them to focus on using AI and robotics and telecoms to let us break free of physical limitations. That really would change what it means to be human.'

'Well, as usual, you've taken everything to the limit. It's wonderful, and I'm sure you'll do it, but be prepared that for most people those last ideas will be *way* too far.'

'You're probably right. In which case, I'd probably better not tell them my extra-wild idea.'

'There's more?'

'It's just that I think I'll be able to turn CHARLIE into a time machine –'

'Oh, for heaven's sake!' He'd snapped to his You're Losing Me expression.

'No, really, it could be amazing! With a bit of work, using my ambiguous reality, I'll have the power to go back in time. It'll all be recorded. And I can play it back. And I won't be able to tell the difference. Maybe I was physically there the first time. Maybe I wasn't. But I can relive it just the same. Better than that, if I want to, I'll be able to *improve* upon the original – turn it into a "best of" experience. Rewrite history by cutting out the rubbish bits.'

Francis was starting to roll his eyes. I intuitively upped my enthusiasm to try to win him over.

'It's even more amazing than that! If the first experience happened to be in VR then, just as Doctor Who encouraged me as a child, I'll be able to go back in time and *change* the outcome. So, my ambiguous reality isn't just an unbounded blurring of physical and virtual, not just past and present, but alternative timelines as well!'

'Now,' he adjudicated as he pressed the ignition, 'you're just being silly.' He reversed, turned and pulled out of the car park. 'But if anyone can do it . . .' We headed homeward across the moor.

'There's just a little more . . .' Francis gave me absolutely no acknowledgment, which I took to be the highest level of encouragement I could expect. 'It's by way of a thought experiment. In the near future, maybe only a few years away, I've just finished giving a speech in New York and I'm being interviewed one-on-one by a journalist. I've just shaken her hand. Meanwhile, I've just finished giving a simultaneous speech in Beijing – identical content but in Chinese. I'm being introduced to the organizer of the conference, who only speaks Chinese, and I'm about to shake *his* hand. All the while, my biological brain is in Torquay. *Where exactly is Peter 2.0?*'

'In Torquay, obviously, like you just said!'

'Yes, and no. I don't think it's quite that simple. In my thought experiment, Peter 2.0 can only do everything I said because he's a lot more than just a biological body remotely controlling a couple of doppelgangers. My voice and the avatar personality that the New York journalist is experiencing are being generated in real time *in* New York. Nothing is originating from Torquay other than a few very high-level directions – no different than a remote movie director shouting 'Action!' to a distant actor who is *improvising*.'

'OK, so are you suggesting that that bit of you is in New York?'

'It is! But exactly the same is true of Beijing – except that the machine translation is probably taking place somewhere in the cloud.'

'Yes, but you'll still *know* that you're actually in Torquay.'

'Maybe, but think about it. I'll *feel* that I'm in both Beijing and New York. The one place I won't feel I am is Torquay. And the one place the people in Beijing and New York will *not* feel me to be is Torquay – even if they know that that's where my biological brain is.'

'So, where do *you* think that you'll be in a situation like that?'

'The true answer is probably that as Peter 2.0 I'll sometimes be in various places at once' – Francis groaned, so I pressed on – '*because* different parts of my persona – which *you* will identify as, in their entirety, being "me" – will be running in different places.' He didn't actually groan at this latest revelation, which I took to be a good sign. 'On those occasions, I truly will be a distributed intelligence. My biological brain will no longer define me, any more than my crippled biological body will.' This was brutal, but true.

Francis said nothing as he focused on bypassing a Dartmoor pony daydreaming at the side of the road.

'I'm not dying, darling; I'm *transforming*. The very fact that the solution to my thought experiment isn't immediately and blatantly obvious is jaw-dropping. It signals a genuine paradigm shift, you know, when *everything* changes, when all the old assumptions are thrown into question. That's what's going to happen very soon. In changing what it means to have MND, we can forever change what it means to be human.'

'When you say it like that, it does sound like a once-in-a-lifetime opportunity.'

'This isn't just a once-in-a-lifetime opportunity – it's *never* occurred before in 13.8 billion years!'

'Are you all ready for tomorrow?'

This non sequitur probably signalled we'd gone as far as we could go in discussing these wild ideas.

'Absolutely! It's going to be a bit of a break.'

'Only you would claim that going into hospital for a major operation was relaxing!'

'I'll get to laze in bed for a couple of weeks – what's not to like?'

'Yes, but remember, you're going to have someone shoving a camera in your face when you're feeling awful and asking, "How do you feel?" and you always click into "showtime mode" whenever you have an audience and you give it all you've got and it drains you of energy. You'll need *all* your energy to get better. I'm not happy about it.'

'It'll be fine. You'll be there. You'll be able to talk to them.'

'What on Earth can *I* say, when the person they want to interview is you?'

'You'll say what you always say when you're protecting me from people who step too far out of line: you'll tell them to bugger off!'

Tripleostomy

Bowel prep is a lot less alluring than I had assumed. I suspect I was misled by its innocuous name. After all, I'd had at least two hours of prep every night at school – ranging from Latin prep to Biology prep – so my brain had no doubt subconsciously pigeonholed this new example into much the same category: 'Your bowel prep for tonight is to read Chapter 5 of *Proctology for Fun*.'

In reality, I was now realizing, it was the means by which a colorectal surgeon reinforced your decision to have a colostomy by ensuring that you would never *ever* want to go to the loo in the conventional way again. In its favour, it was the ultimate example of going out with a bang.

It had all started very humanely. Around midday, Francis and I – complete with TV-documentary film crew – arrived at the main entrance of Torbay's NHS hospital. There we were met by a media wrangler kindly provided by the management. My growing entourage proceeded to make its way to the appropriate floor and eventually arrived at the entrance to the ward. Then we arrived again, so that the camera crew could get an establishing shot. It was only when we arrived for the third time (filmed from the opposite direction) that we were allowed to check in to my new lodgings.

Immediately, things went slightly off plan. Francis and I were escorted away by a woman who appeared to be in charge of the woman we'd thought was in charge. Confusingly, she wanted an unscheduled meeting 'in camera' (which, given that I was with a camera crew, and given that 'in camera' is Latin for 'in secret' and therefore meant '*not* on camera', was a confusing choice of words). As it turned out, our in-camera-not-on-camera meeting was relatively short. She passionately argued that my proposed operation was too risky to be allowed to proceed.

'I've seen people with MND completely destroyed by having just one operation, and you're effectively having three. We'll totally support you if you still want to go ahead, but you really need to understand that it will almost certainly be a very long road to recovery. And you may *never* fully recover.'

We were both already well aware of the risk. As ever on major issues, having discussed everything to death (including death), we were both also of one mind. In a united counter-attack, Francis and I efficiently overwhelmed this unexpected last-minute opposition with a crisp but robust explanation of what 'long-term overall quality of life' actually meant, said thank you, and the meeting ended, somewhat earlier than our host appeared to have expected.

We rejoined our entourage and a brand-new woman, who appeared to be in charge for the moment, took us to one of the small rooms off the main ward. This was to be my holiday home as soon as I returned from a short stay in Intensive Care. My room upgrade was, I'd learned, less a perk of travelling with a film crew (I'd been allocated the room before the hospital realized I came with excess baggage) and more a benefit of my NHS computer record flagging up that I was on the 'gold standard' – a classification of which, thanks no doubt to growing up in Wimbledon, I'd felt unjustifiably proud until discovering that it meant I was on end-of-life care.

A couple of hours later, my initial TV interviews were over (yes, I really did feel fine; no, I really didn't mind hospitals), the film crew had departed and Francis had just kissed me goodbye and left for the day, having first helped me into a rather fetching hospital gown and on to a commode – which is where he had been instructed to leave me. In the absence of anything to do, I caught up with the news on my mobile.

'I've been told to give you bowel prep.'

The healthcare assistant seemed very pleasant, and smiled encouragingly as she placed a large plastic jug of slightly tinted water on the side table next to me. I knew that for my colostomy the next day they'd want my colon as empty as possible, so I'd half expected an enema. Just drinking something seemed a far more attractive proposition.

'What exactly do you want me to do?'

'Well, I'll just throw the water away from your glass and then if you could drink some of this . . .'

It tasted a bit weird. She encouraged me to take another sip. It still tasted weird. She now urged me to finish the whole glass. Halfway through, I decided that the taste was sufficiently unpleasant that continuing with decorous sips was inappropriate. I necked it.

'It's not the most pleasant taste in the world, is it?' I ventured, replacing the now thankfully empty glass on the table.

'No, I'm afraid not. If you can make sure you finish the whole jug within the next half hour, then it should start working in about an hour's time.'

With that, she smiled and left, leaving me and the jug staring at each other.

Half an hour later, I had increasingly queasily done my duty. I didn't feel great, but I felt great that the jug was now standing empty.

'Oh, well done! You've finished it.' She was as smiley as ever as she swept up the jug . . . and replaced it with its completely full twin. 'You just need to get through this one and then you can relax.'

An hour later, I was anything but relaxed. I'd reluctantly forced down the contents of the second jug. And then I'd waited. And waited. Nothing changed other than that my queasiness increased. Then the woman in charge of marking my belly with a felt-tip pen to show where my rerouted colon should emerge returned and asked if I was ready for her. I'd completely forgotten that earlier I'd asked if she could give me a bit more time. I asked her again, possibly a little more urgently than before. She said she'd come back early tomorrow morning.

Then it started. First an unexpected dribble. Then a deluge. Then nothing. There wasn't a lot to do other than sit and wait. I began to understand why in medical terminology one 'evacuates' one's bowels. That was exactly what it felt like: I was definitely being evacuated. Another deluge. Nothing. A torrent. Then my mobile started ringing. I looked at the caller ID. It was the chair of the MND Association who'd very kindly offered to ring and let me know whether I'd been elected as a trustee. I was between bouts so decided to risk answering.

This decision brought mixed blessings. On a positive note (and to my great surprise), the extended membership of the Association had elected a total stranger over far-better-known candidates, purely on my ticket of thriving with technology. On a less-positive note, Alun, the outgoing chair, was in congratulatory mood. After all, I'd beaten the odds. And I had a revolutionary message. Why wouldn't he want to discuss it? And share some words of wisdom. And ask about my plans. And ask about Francis. And tell me about himself.

Which was all wonderful, and I really warmed to the guy and I certainly didn't want to curtail the conversation, let alone end up sounding brusque. But I was also acutely aware that my mobile had an especially sensitive microphone. It felt like basic phone etiquette to try to avoid any extraneous noises, but this became increasingly difficult, and my voice must have progressively sounded more and more inexplicably strained. The call and my bowel control ended at much the same moment.

Early the next morning, after the optimal location of output no. 2 had finally been marked with felt-tip, an orderly arrived to relieve me of any jewellery. Apart from my watch, I only ever wore my wedding ring and a gold ankh on a thin gold chain around my neck. The ankh was the ancient Egyptian symbol of life – just like Rahylan had won at the Warlock Games – and Francis's gift to me on our twenty-fifth anniversary. I hadn't taken off either my ring or my ankh since Francis had first put them on me.

For scientifically silly but romantically obvious reasons, I now found I didn't want to remove either of them. Despite all my training and all my beliefs, and lack of beliefs, a primitive part of my brain wanted to believe that such gifts, given in love, carried some ancient power. Or maybe it was the part of my brain that was forever with Avalon in Salania, where Rahylan's magic was unequalled. Either way, I just didn't want to give them up.

I pulled myself together, reminded myself I was a scientist and handed over my ankh. And lied that my wedding ring was too tight to get off my finger. We'd need to tape it up. Which we did.

The film crew did a quick pre-op interview. I made the point of saying that, even in the unlikely event that something went wrong, science would benefit because we'd have learned from the failure, and that was what research was all about. Then they filmed Francis and me kissing goodbye, and I was wheeled to theatre. I had a lovely chat with Maree, my anaesthetist, as she inserted an arterial cannula into my wrist, a few pre-op tests to confirm that despite my lungs now only working at 76 per cent there were no surprises, a reminder from me that they should keep going regardless, and without further ado she knocked me out.

Francis waited, like all loved ones wait, with nothing to do but wait. Then he got a call. I was being taken to ICU. He got there shortly after I did. The difference from normal hospital protocol was that, as planned by Maree, I would still be fully anaesthetized with propofol and remifentanil and would come off them one at a time – in case I needed to be rushed back to surgery to have a tracheostomy.

I came off the propofol and they waited till it was out of my system. Then they turned off the remifentanil. Now, the thing about remifentanil is that you ought to recover from it in ten minutes or so. So as soon as the mechanical ventilator was only feeding me air, the countdown was on. At almost exactly the ten-minute mark, I regained consciousness. The ICU sister told Francis it was the fastest she'd ever seen any-one recover. But that, of course, was not the point. The big question weighing on Francis and everyone else in the room was whether I would ever be able to breathe on my own again.

My very first memory after coming to was seeing Francis and being told the operation had gone well. I was in a bright room, very modern, people around me. I was wearing a big mask, a bit like a gas mask, strapped around my head. I didn't like the way it was fighting the way that *I* was trying to breathe. I indicated that I wanted to experiment with taking it off.

For some reason, Francis decided that this of all times was a good moment to start filming proceedings on his mobile.

And for some reason, the authority figures in the room let him. As a result, he caught for posterity the moment when they took off the mask.

And I took a deep breath on my own.

In the middle of the night, the young nurse who had been monitoring me since she had first come on shift at 8 p.m. brought my meds. We'd been chatting for hours, about my tripleostomy, my MND, my epiglottis episode, my plans to thrive, my relationship with Francis. I felt wide awake. All but one of my meds were in tablet form and I knocked them back with some water, grateful that my ability to swallow appeared totally unaffected by the op. The pharmacist had provided the final med in liquid form – which, going forward, made more sense than risking choking on tablets. I drank from the disposable mini cup. It was a syrup. It stung a little as I swallowed.

Then it stung a lot. I started to cough uncontrollably. Very quickly, far faster than on Dartmoor, the stridor began. I was whooping, then a quick cough, then a whoop again. Far too little time to draw a breath. I could feel myself asphyxiating. My lungs, despite me being full of painkillers, burned as they screamed for air. My nurse was caught off guard. Was I choking? Was it an allergic reaction?

In my desperation, I gambled what felt like the last air in my lungs to urgently hiss:

'Epiglottis!'

Two weeks later, I was cradling my brand-new great-nephew, Eddie, who had very graciously overstayed his welcome in his mother's womb until the day I was allowed home. Eddie was allowed home too, and now he and his entourage had popped next door to see us.

'And did she understand what you meant?' one of my in-laws asked urgently, as if she needed the answer to keep me alive.

'She looked confused for a couple of seconds and then she remembered my story. I guess that had been the last time she'd heard the word "epiglottis", so it was easy really. Anyway, she grabbed some water, got me to drink and the rest is history!'

'It could have been awful!' another in-law chipped in.

'Yes, but honestly, if you're going to be asphyxiated, the place to do it is ICU.'

'So now that you've paved the way, does that mean that other people can have the same triple-thingy on the NHS?'

'Absolutely! Of course, the risks remain. I was very lucky to do so well, but now we've proved it's a valid option for MND.'

'Oh, tell them what Tracy said,' prompted Francis.

'Ah, yes, well, our lovely friend Tracy is responsible for coordinating all the NHS care for people with MND in the South West, right. So, she tries to go to people's homes soon after they're first diagnosed. Well, evidently the other day she went along for her first meeting with this little old lady, and after they'd got over their initial pleasantries, the lady says: "Now tell me, my dear, have you ever heard of a man called Peter Scott-Morgan?" Tracy, naturally, is now on high alert and replies: "I've *heard* of him . . ." "Excellent! In that case, in addition to a feeding tube, I want something called a supra-pubic catheter." Tracy was *so* proud!'

Berkeley Square

I checked the Cartier watch that Francis had recently given me for my thirtieth birthday – almost half-past seven – buttoned up the waistcoat of my dark-navy three-piece suit, glanced in the mirror, picked up my briefcase, added to its contents the pack of freshly made sandwiches that Francis handed me, gave him a big kiss and headed at a brisk pace to the nearby Osterley Tube Station exactly on schedule – little knowing that my day was about to spiral out of control.

Osterley, we'd discovered, was ideally located for my current lifestyle; from there it took about the same time by Tube to travel east to my Mayfair office or west to Heathrow, and some weeks it was a toss-up which direction I took most. Another advantage of living forty-five minutes from the centre of London was that most mornings, provided I was early enough, I got a seat on the train – a few stops later and people began having to stand.

Today, I took the eastbound train and spent the trip running through my presentation. I entered the imposing entrance of Berkeley Square House almost an hour later, called one of the marble-clad lifts and pressed the button for the ninth floor – the penthouse. This was where the misfits like me were kept.

In a bizarre twist of office layout, the London offices of the worldwide technology and management consultancy ADL – Arthur D. Little – were on two floors of BSH, the sixth and the ninth. The sixth floor was huge (and therefore the centre of power), whereas the ninth floor was a smaller extension added on to the roof by an industrious landlord when office rentals skyrocketed. Despite the fact that the plate-glass windows of my office had panoramic views over Central London, none of my colleagues in their smaller, un-air-conditioned offices on the sixth floor would ever have swapped with me because *they* were close to the Power Corridor that ended at the corner office occupied by Michael (we were all referred to by our first names), the managing director of the London office. I was happy where I was and didn't particularly want to swap.

The Ninth-floor Gang rather luxuriated in being rebels, outcasts from the mainstream. The very fact that we had offices at all – let alone nice offices – meant that ADL was making good money from us and therefore wanted to keep us around. And happy. Just not too close. I was by far the youngest on the floor and, having been with ADL for only four years, was also the most junior. That said, most employees only got to stay for two years before they were politely encouraged to move on, so I must have at least been viewed as 'promising'. Perhaps linked to that, I was also the only one in the office with a PhD.

I shouted a greeting to Robert through the half-glazed door to his office. He was about twenty years older than me, smoked like a chimney and was possibly the wildest occupant of the floor. The stories about him were legion, and most involved women, drink, drugs or a heady combination of all three. But his clients loved him and, in the style of a forgivingly forgetful uncle for the hundredth time admonishing a favourite nephew after yet another precocious transgression, ADL loved him too.

So, by all accounts, did some of the secretaries. Frequently. I knew this to have been true in at least one instance. About a year earlier, as I was returning to my office, a colleague standing outside Robert's door had seen me approaching and urgently beckoned me to hurry to his side.

'He's screwing that new secretary from the sixth floor! I saw them go in ten minutes ago.'

Sure enough, from the sounds emanating from the office beyond, his conclusion might well be true. However, we would never know for sure because the window in the door was covered by what looked like a very large sheet of shiny white paper. I enquired what was obscuring our view.

'It's that poster he has on his wall. He peeled it off and Blu-Tacked it over the glass to get some privacy.'

We'd been about to leave when we noticed something: excellent though Blu Tack is, it is a sad fact that when it's old and dried out by ceaseless air-conditioning it can lose some of its adhesive properties. And so it was now. One corner of the poster was no longer attached. As we watched, the other corner sprang free. Like a blind being slowly lowered, the poster gradually curled downwards until the Blu Tack at the bottom gave up completely and the poster slipped to the floor, revealing that my colleague's earlier inference regarding the sound effects had been totally accurate. Robert, no doubt disturbed by the errant poster, looked up and saw us. True to form, he simply carried on, totally unperturbed. His only concession to having an audience had been to smile and give us a thumbs up.

In the late eighties, ADL was an extraordinary institution. It had deep roots in technology, but also in how to innovate, how to make technology companies successful, how to run them, how to use technology to change things forever. That combination alone had enthralled me soon after I joined, and I found that I loved the consulting lifestyle. But I was also, I rapidly discovered, supremely lucky to have joined a corporate culture in which, with a bit of care, I could flourish.

ADL maintained a century-old love affair with freethinking characters like its larger-than-life founder, Dr Arthur Dehon Little himself. This was the company that had used its skills in chemical synthesis to literally make a silk purse (actually two) out of a sow's ear (actually a cartload of pigs' ears). This was the company that, not to rest on its laurels, a generation later literally made a lead balloon fly (*so* high that it became an aviation hazard at Boston's Logan Airport). This was the company that had devised such a clever way of categorizing smells that Ernest Crocker, the 'man with the million-dollar nose', had saved a boy's life by sniffing out what was wrong with him after his doctors had given up hope of a diagnosis. This was the company that I hoped would be sufficiently supportive of my new idea that they would let me develop it.

Just before lunch, I ate my sandwiches. Everyone else I knew bought theirs from one of the multiple vendors nearby, but Francis and I had calculated just how much we would save by me taking in my own. We both vividly remembered what it was like to feel poor and neither of us wished to repeat the experience. While all my colleagues seemed to live to their new-found means, Francis and I saved as much as we could to keep paying off lumps of our mortgage.

As it was, today I was about to go to the conference room on the sixth floor, where a whole selection of sandwiches would be available for the monthly meeting. However, it was I who was due to give the presentation; by the time I'd finished, nothing worthwhile, and certainly nothing that hadn't been fingered multiple times, would be left.

Even forty minutes into my presentation, I noticed that all the food was gone. With their mouths no longer stuffed with bacon and avocado or prawn or salmon and cottage cheese, they began to ask questions. Eventually, Patrice from the

Brussels office took the floor. He was a vice president and one of the grand old men of ADL. He was also head of the global technology innovation practice and therefore my ultimate boss. He'd flown in for the day. Now, for some reason, he'd stood up to ask his question; no one else had stood to ask theirs. He was very tall and sported a very precise beard, with a very Belgian accent to match.

'So, Peter, this Unwritten Rules technique of yours, I understand it's essentially a technique for decoding corporate culture?'

'Well, that's a by-product. But it's really a way of decoding hugely complex systems – even organizations of tens of thousands of people – so we can explain their hidden inner workings.'

'Yes, but what you are focusing on is the soft, fluffy culture, no?'

'Not really, no. The whole analysis technique works by imposing rigorous logic at every step. Yes, you're right, in the process it also explains aspects of what otherwise gets dismissed as "soft and fluffy" – but that's an added bonus. The point is, with this technique we can at last explain *why* and *how* some of our Great Ideas fail. We can then refine the ideas so they actually *fit* with an organization, rather than simply trying to force them into an organization that rejects them.'

'*That*, my friend, is exactly the soft and fluffy culture that I was afraid you were talking about.'

He looked around his audience as if he were an expensive barrister addressing a jury. This was not going well.

'But Patrice, at its heart, the Unwritten Rules technique is *not* about corporate culture, it's about decoding strands of the future. It's about removing hidden barriers to change. Ultimately, it's about changing the future. And, although they don't tend to use those words, changing the future is ultimately what our clients pay us to help them do!'

'Yes, but not to change their culture!' It was as if he wasn't listening. 'This is not an issue that concerns any of our clients.'

'But it does! We keep giving them intellectually neat solutions that don't actually work in the real world. You know very well that's how I first came up with the technique, analysing at Philips Electronics why some of our recommendations weren't working.'

This was possibly not the most diplomatic response I could have given – they'd been Patrice's recommendations that had failed.

'Philips do *not* need a "culture audit" or "climate index" by another name. I'm sorry, but it seems to me that you're *whipping up candyfloss out of nothing.*'

Once again, he addressed the members of the jury. 'Ultimately, whether your Unwritten Rules technique does or does not work, it is *not* what Arthur D. Little should be doing.'

With that, he sat down, Michael, the managing director, stood up and thanked me as 'our resident maverick' for a 'stimulating' presentation, and the meeting broke up. I gathered my overhead-projector slides and was about to head back to the ninth floor when my mentor, Bruce, walked over. He too was a grand old man, a VP, in all but name the deputy MD. He was approaching retirement, notorious for his irascibility, a pedantic perfectionist, a scourge of unstructured writing, living with a formidable lady who had been the first ever female partner at the prestigious consultancy firm McKinsey, and I liked and respected him immensely.

'A quick debrief?' he murmured.

We walked to his office, next to Michael's, me trying to match his long, slow, calm strides so as not to appear frantic in comparison. He waved me to one of the chairs and sat down himself.

'Don't pay undue attention to Patrice. He's brilliant, but he can't conceive that anyone else could be *as* brilliant, let alone risk anyone putting him in the shade.'

'But surely, as I'm part of his department, Unwritten Rules would only boost his credibility.'

'Ah, but he wasn't the one to come up with the idea, you see. So in his mind, it can't possibly be worth consideration.'

'I'm sorry but if that's true, then it's ridiculous. He's bullying me – and I don't give in to bullies!'

'Calm down, calm down, no need to come out all guns blazing. Try and sell your idea to some clients. There is nothing more persuasive than bringing in money!'

'But he's blocking me from all of his clients.'

'So go round him, sell it elsewhere. You've got that proposal to British Petroleum already. How's that going?'

'I've heard nothing. It's been ages. My contact at BP says it's still in play but they're "discussing scope" and there are lots of people who have to sign anything off.'

'OK, so your contact may be letting you down gently, but you never know. You *will* get a big sale one day. And that will consolidate your position in the firm, not just as a creative "maverick"' – Bruce smiled as he referenced Michael's earlier description of me, but he didn't distance himself from it either – 'but as a potential junior partner.'

He'd confidentially mentioned to me a few weeks earlier that – along with Jarvis, a slightly older colleague – my name had been accepted for consideration at the annual vote in a couple of months' time. I had an outside chance at best.

'Well, I'm not holding my breath,' I laughed.

'As I said before, if you got it this round, then you'd be the youngest ever junior partner, so we shouldn't get too carried away. But I wouldn't have proposed you in the first place if I didn't think you were in with a chance.'

'I'm really grateful.'

'It's my job to ensure that the right senior consultants get promoted.' He paused and seemed to be choosing his next words with care; both behaviours were unusual for Bruce. 'That's why I'm not very happy with what's being said about your private life.'

Unwritten Rules

Suddenly, it was my 'little chat' with the headmaster all over again. Bruce and I had *never* discussed my private life. Then again, even more so than at school, I was proudly gay and said so to any of my colleagues who chose to ask. Bruce had simply never asked, any more than Michael had. But both of their secretaries were friends of mine (especially Helen, Michael's PA) and by now those two knew intimate details of my private life, so I'd always assumed that their bosses had been given partial synopses that they'd chosen never to mention. Until now, I'd also always assumed that they were OK with it.

'This is just between you and me. I do *not* want you to do anything about it. But it has come to my attention that Jarvis is having conversations with the partners about your suitability for promotion.'

'What?'

Bruce gave an avuncular wave of his hand, telling me to calm down.

'He seems to have got it into his head that only one of you can be promoted to junior partner – and he doesn't intend it to be you.'

To my surprise, I realized I wasn't surprised. Jarvis had always struck me as a cut-throat street trader who'd had elocution lessons and become a cut-throat bond trader at the London Stock Exchange.

'What, exactly, has he been saying about my "private life"?'

'Oh, nothing bad, nothing attributable. Just innuendo. Concerns. You know the sort of thing.'

'No, I don't.' Of course I could guess.

'Things like: "I know it's a problem with me rather than with Peter, but I struggle with his lifestyle. And I know it's his private life, and nothing to do with his work, and he's incredibly talented and a great asset to the firm, of course, but I do worry how our *clients* might react. After all, some of them are very conservative. So, however *we* feel about Peter's life choice, we really have to put our *clients'* feelings first." For Jarvis, it's actually rather clever.' Bruce didn't rate Jarvis's IQ. 'He ends up sounding like Mother Teresa as he sticks a poisoned dagger in your back!'

Very clever. Bruce and I agreed there was nothing I could do – other than continue to try to excel at my job – and with that the debrief was over and I returned to the ninth floor, well aware that Jarvis's office was on the sixth.

I spent the next two hours in slightly deflated mode, finalizing a client presentation I was scheduled to give in Dublin the following day. My secretary excitedly popped her head around the door and, as she placed some papers into my in tray, imparted her news:

'The fax room just rang up and said a letter has been faxed in for you. So it must be urgent. But the office boy is out buying some fax paper, because it's just run out, so you were lucky the fax got through; they said it started beeping as soon as your letter got printed. Anyway, it's all safe but no one can bring it up because there's only one of them there at the moment, so do you want me to go down or can it wait? It's just I'm still finishing those final corrections for you, and I've got to leave in less than an hour because I'm meeting that Italian guy again and it'll take me at least half an hour to get ready once I get home and –'

'Did they mention who the letter was from?'

'Oh yes, it's from BP.'

I was already up from my desk and sidling past her into the corridor as I reassured her, 'Don't worry, I'll get it myself.'

I legged it to the lift but rather than wait I leaped down the emergency stairs three at a time and at the sixth floor bounded into the reception.

'Busy, busy boy!' chirped one of the immaculate young receptionists.

'Man on a mission!' chimed in the other. I beamed back a smile as I slowed my run to a fast walk. Around a corner into the main corridor, off into a short side corridor and into an office-sized room with a telex machine, photocopier, two fax machines and a middle-aged woman.

'I believe you have a fax for me?'

'Yes, dear, here it is.'

I took the shiny fax paper from her. She'd already cut the continuous thermal paper into two roughly equal sheets to match the original two-page letter. Coming from the end of the roll, the paper was tightly curled and I could only read a paragraph at a time. Having got to the end, I laid the two sheets flat on a tabletop, held them down to stop them curling and read the whole letter again.

'Is it good news, dear?'

I turned to her triumphantly.

'The world is mine!'

I snatched up the curly fax and marched down the Power Corridor to Bruce's office. My triumphal march would have ended in something of an anticlimax if his door had been closed. Fortunately, it was wide open and, seeing me coming, he waved me in without me having to break stride. Without either of us saying a word, I handed over the fax. He read it quickly. By the end of the first page, he was smiling. At the end of the second, he passed judgement:

'*This* changes everything!'

He bounced up from his desk and strode as fast as I'd ever seen him stride across the corridor to Helen's office.

'Is he free?'

'He's deep in conference with Patrice, I'm afraid.'

'All the better!'

Bruce gave a perfunctory rap on Michael's door, didn't wait for an acknowledgment and burst in.

'I knew you'd want to see this,' he offered as an explanation for our explosive arrival. Patrice looked irritated, then resigned to the interruption, then irritated again when he saw me in Bruce's wake. Michael, in contrast, saw me and smiled an intrigued welcome.

'What's this then?'

He took the sheets, spread them on his expansive desktop, pinned them down with sundry items and began reading.

'Aha!'

He must have reached the place where the letter explained that, instead of having an Unwritten Rules analysis of *one* of their sites (as I'd hoped), BP had decided that they in fact wanted appraisals of *ten* of their sites.

'Oh, well done *you*!'

He looked up and smiled at me before looking down again to polish off the end of the letter. He must have got to the bit that spelled out just how much BP were committing to pay – the bit with a number followed by a gratifying number of zeros. Plus expenses.

Patrice could not contain his curiosity any longer.

'Might I ask what is the good news?'

Bruce couldn't resist the opportunity to crow.

'Well, Patrice, with reference to your comments at lunch

today, I am delighted to inform you that young Peter here has found a very knowledgeable – and very rich – client who appears to prefer candyfloss over Belgian chocolate.'

When Patrice looked bemused, Bruce added:

'A *lot* of candyfloss!'

Weeks later, Francis and I clinked our flutes of champagne and Francis offered a toast:

'To the youngest ever junior partner!'

'You know, they're moving me down to the sixth floor next week – only four offices away from Michael.' Francis looked suitably impressed. 'It just happened to come available,' I admitted, 'so it doesn't really mean anything. But nevertheless, Jarvis's office is miles away!'

'Good!' He took another sip. 'And what does this all mean for your Unwritten Rules idea? What does it mean generally?'

'Well, people will take me a little more seriously for a start. So long as I pay my way, I'll tend to be accommodated. And now I'm a junior partner, I can do my own thing a little bit more than before.'

'You've *always* done your own thing!'

Over the next two years, we had two more key moments to celebrate. Our twelfth anniversary also marked the first day that a legal change to our surnames came into effect – we were now the Scott-Morgans, and we couldn't have been happier. I had worried about how my parents might handle it, but they took it in their stride. Ma actually said how very proud we should be of what we'd achieved, standing up to bigotry – it showed just how far she and Da had travelled towards accepting and even celebrating our relationship. And although she didn't express it directly to me, she made a point of saying to Francis how guilty she felt about their treatment of us earlier on, how brave we'd been and how Francis had been the best thing that ever happened to me – I couldn't have agreed with her more.

The second occasion was no less momentous, as it meant a wholesale change to how and where we lived. I'd spent a week at ADL's corporate headquarters, in Boston, USA, having been asked to give a presentation on Unwritten Rules to anyone who was interested. Naturally, I'd obliged; it seemed to go well and I'd thought nothing more of it. Then, during a meeting at the end of the week with the managing director in charge of all North American management consulting, she'd casually said, 'I'd like to send a team over to London for six months, to learn from you all about how to conduct Unwritten Rules analyses.'

I hadn't expected this. But within a second my brain made a life-changing decision. It felt *so* right that without taking any time to consciously consider it I immediately answered:

'No, no, it would be far more efficient if I moved over to the USA for six months. That way I can train everybody you want me to.'

She'd agreed on the spot.

When I'd rung Francis, he'd instantly agreed as well – with the obvious stipulation that he would come with me. When we'd discussed it in depth upon my return home, he'd been unequivocal:

'This is your big chance. We've got to give it our all. Things will never take off in a satellite office like London. But in the USA, who knows what could happen? We don't want to spend the rest of our lives thinking, "What if . . . ?" This is the time for us to go for it!'

Based on that, he'd handed in his notice, we'd put our house on the market, arranged to farm out our two Shetland sheepdogs to alternating sets of parents and prepared to move to Boston, Massachusetts. Our grand plan was to make our fortunes in the USA and stay there indefinitely. That was our dream.

Our reality was that Francis had never actually visited Boston, did not have a US visa for anything other than a vacation and, as far as ADL was concerned, I was only on a temporary executive transfer. The email regarding my move to the States was very explicit: it was for a maximum of six months.

The American Dream

Rather inauspiciously, Francis and I arrived in Massachusetts during a state of emergency; snow had shut down everything. Unperturbed, we asked our taxi to drop us at 'the centre of Boston' and from the snow-encrusted golden dome of the State House we trudged along the even more snow-covered sidewalk. Rather auspiciously, the first apartment building we came to had a studio that had just become available, on a high floor with a glorious city view. We grabbed it. Our search for rental accommodation had taken us less than an hour and less than a hundred paces from the State House. We'd arrived.

The centre of Boston turned out to be an amazing place to live. So much so that we were soon toying with the idea of buying a home; we had no intention of being repatriated to the UK on schedule. To that end, from the moment we first set foot on American soil I assiduously struggled to crowbar open the rigid six-month window of opportunity ADL had given me. It had cracked open in stages.

Despite us having moved to the USA on the understanding that it was only a short-term arrangement, soon after my arrival I'd been introduced as 'Peter, who's here for the next year', and when my official transfer documentation arrived a few months later, it clearly stated that I would stay in the USA 'for a minimum of two years, to be extended by mutual consent'.

So it was, six months into living the American dream, with my window of opportunity now wide open and my dreams commensurately unbounded, I was ushered into the expansive top-floor office of the head of corporate marketing at ADL's headquarters.

'Come in, come in, come in!' He looked to be in his forties, had an educated East Coast accent and dressed accordingly. 'So, it's all over, eh?'

As I sat in an uncomfortable designer chair across the coffee table between us, I appreciated the view from his window across the large Acorn Park campus dotted with laboratories, offices, helicopter pad and lake complete with Canada geese. My office, in a connecting building, hardly had a window at all. I also appreciated his frigid air-conditioning, which brought the temperature down almost to that of a New England winter. Today, in contrast, outside it was beyond hot and incredibly humid.

'So, Peter, as I said, the presidents' dinners are all over and I guess you want to debrief?'

The presidents' dinners, eight of them, in major cities across the States, had been a big deal. ADL only put them on once a year, and only chief executives were invited. For the 1993 series, I'd been the only speaker – forty minutes on 'The Unwritten Rules of the Game', strategically placed between the main course and dessert. Indeed, it had been such a big deal that when Francis and I had taken a week's vacation in Key West during the worst of Boston's February snow, ADL had kindly flown two people down to work with me on my speech (in T-shirts beside a pool) rather than fly me back to Boston early.

'Actually, I've an idea of how we can build on the success of the presidents' dinners.'

'Well, it's our single biggest marketing investment of the year so the more we can leverage it the better. What do you have in mind?'

'I want to write a book.'

His face froze. Not a good sign.

'Aah!' An even worse sign. 'I'm not sure.' A positively bad sign. 'You see, there are some ideas that are perfectly adequate for an article in a magazine, or a speech – like you so excellently delivered at the presidents' dinners – but . . .' the combination of the hanging 'but' and the pause as he chose his next words was an excruciatingly bad sign, '. . . that same idea would be stretched a little *thin* if forced to fill a whole book.'

I spent the next half hour attempting to convince him that I had a *great* book in me.

'I just need some time to write it,' I concluded and locked eyes with him, hoping what I'd said had done the trick.

'I'm sorry but I'm as unconvinced as ever. We can't support you, I'm afraid. I simply don't see this as an ADL book. Sorry.'

A few minutes later I was crossing the elevated walkway from Corporate back to North American Management Consulting. The walls transitioned into painted cinder blocks – allegedly to make it feel like the labs of the nearby Massachusetts Institute of Technology.

'Hi, UROG Pioneer!'

It was one of the high-selling VPs, crossing the bridge in the opposite direction. He always smiled when we met. Then again, he always seemed to see me as something of a joke. In our first ever meeting, he'd blithely turned to a colleague and said: 'He's into fags, you know.' Surprised, but unfazed, I'd innocently responded: 'Actually, I don't smoke at all. But I'm sure I could get you a fag if you're desperate for one.' He'd smiled, and I'd sensed I'd passed some sort of test.

'Hi!' I let slip his use of the awful abbreviation of 'Unwritten Rules of the Game' to 'you-rog'. At least he'd called me a pioneer, which was flattering. As I headed down some stairs, I reflected that I was now a definite step up from maverick.

After a few more corridors, I made it to what was amusingly referred to as my office. It had been allocated when I was only expected to stay for a few months; it was smaller than the janitor's closet and had a tiny window overlooking the car park. Its location was also a source of general hilarity. As my appointed head of department had joked to my new colleagues when first introducing me:

'Knowing Peter's lifestyle, we've carefully arranged for his office to be right next to the restroom.' He'd turned to me, laughing at his own joke. 'Which we're sure he'll appreciate.'

Months later, I'd learned that it was a general assumption amongst my colleagues that gay men spent any spare time loitering in lavatories. At the time, I'd merely replied: '*Despite* my office location, it's wonderful to be here!'

I sat down in front of the dented metal desk that, along with me, largely filled the room. As if by magic, my loyal secretary, Elaine, appeared at the door and stood; after all, there wasn't much room for her to come in. I could tell she was about to leave because her elegant shoes had been replaced by sneakers.

'How did it go?' After I'd given her a synopsis, she leaned towards me conspiratorially. She'd worked at Acorn Park for so many decades that she knew all the gossip since the sixties. 'It's no surprise. In his last company they said that "good ideas go to his office to die"! My Peter will work it out . . .'

An hour later, I collected my briefcase and the rather ancient laptop I'd been loaned and, despite the summer humidity, briskly walked the mile to 'the T', caught a train, eventually emerged near the State House and minutes later was in our tiny apartment telling Francis all about my meeting, with an American cityscape outside the windows that even after six months still had me expecting to see Superman flying past.

'You'll just have to write it anyway!'

'When?'

'At weekends.'

So I did. Five (*very* long) weekends later, I presented my managing director with a near-final draft of *The Unwritten Rules of the Game*. I'd just hammered it out, drawing on every speech, training course and vague idea from the previous few years. I assumed that the whole thing would now be torn apart by editor after editor, so I'd just written it as I spoke.

I have a theory that prior to sending it to the publisher no one actually read it – everyone took for granted that someone else had. Whatever the truth of the matter, by the time that McGraw-Hill published it as the lead book of their spring 1994 catalogue, only five words had been changed. And two of those were 'fuck'.

Having been elevated from maverick to pioneer, some time during 1994 my inexplicable rise took me into the stratosphere as the press (and corporations willing to spend real money) dubbed me a guru. I was signed up by an international speakers' bureau and also made a full partner. Suddenly, I was a potential revenue-generator. If my colleagues hadn't read my book already, they certainly did after I spoke at the World Economic Forum in Davos. And then they invited me to fly to whatever country they were in (and ADL had a foothold in most places around the world), only to find I was booked solid for the remainder of the year but could slot them in around June '95 . . . if I felt like it.

Francis and I celebrated by buying a Louis XIV ballroom and a muscle car. To be more precise, we fell in love with and took out a huge mortgage on a large condo in an 1860s brownstone townhouse on Commonwealth Avenue, the leafy boulevard close to Boston Common in the very centre of town. The main room was the enormous rococo ballroom of the original house, panelled with hugely intricate carvings taken from a chateau in France. We were possibly overcompensating for the still-vivid memory of sticking Sellotape around the windows of our first home together.

Our second purchase was far more restrained. We knew the feel of automobile we wanted, we just didn't know what to ask for. I tracked down a senior consultant at Acorn Park who happened to be a world expert on the US auto industry. I explained my predicament.

'Ah, what you're describing is a muscle car!'

He gave me some names, we got some catalogues and decided on a Pontiac Firebird. Francis rang a dealer in Greater Boston.

'Yes, we'd like it with the removable roof. Yes, all in the brightest red. No, we don't need a test drive. No, we've never driven one – in fact, I'm not sure we've ever even seen one. Yes, it looks great in the catalogue. And I'm sure it comes with an engine.'

As we purred down Comm Ave, roof off, multi-speakers blaring from the auto-change CD player, wraparound sunglasses protecting us from the summer glare, we were also celebrating a minor victory regarding US Immigration.

Ever since we'd moved to the States, Francis had been at risk of being refused entry. I was on an executive transfer; officially, he was on vacation. The law said that the unmarried partner of someone on executive transfer could travel in and out of the USA as much as they liked, but Francis wasn't recognized as my unmarried partner. This seemed unfair. We worked with an immigration attorney to make our case, and eventually an official clarification of the law arrived that agreed that 'gender is immaterial' when it came to defining an unmarried partner. Francis flew into Logan Airport brandishing his newly minted right of entry; this set legal precedent and was communicated to every immigration attorney in the country. Job done.

The competitive dynamic created across ADL by the need to lure me to visit one far-flung country rather than another led the offices around the globe to attempt to outdo each other in the creativity they displayed in attracting their clients (and me, often with Francis by my side if he liked the location) to attend an Unwritten Rules extravaganza.

Within a few months, this became my life. It was exhilarating – but tiring. Everyone assumed I was a workaholic extrovert who was now in his element. In reality, whenever Francis and I grabbed a few days' break I never had the slightest difficulty in not checking my voicemail or email. I worked hard because that's what I'd been trained to do at school. And every psychological profile I'd ever taken showed me as an introvert who had learned to fake it. If ever I got the chance, I used work as an excuse not to dine with colleagues and instead escaped to my hotel room and ordered room service. And worked.

After two years, I was beginning to learn the Unwritten Rules of being on 'the circuit'. But I was also being run ragged. I'd just gone around the world in eight days – west to east, which makes the jetlag worse – and given seven speeches en route. By the last leg, I was spending more time on autopilot than the planes. My only memory of the Far East was introducing myself to a journalist with the words: 'Peter Scott-Morgan,' and her looking very disappointed.

'Couldn't he come?'

When I'd explained that he could come, and that I was he, she'd burst into a smile.

'Oh, I thought you'd be fat and old!'

Other than that, all of the exotic Orient was a blur. Which, as Francis pointed out, might have been an acceptable trade-off if there'd been any upside for us. As it was, all book royalties and all speaking fees went to ADL. Worse, there was now a new MD who had a relentlessly one-size-fits-all view of management; and by any stretch of the imagination, I did not fit. Nor, frankly, with him in the role, did I *want* to fit. Francis, as ever, supported my rebellion.

'I'm going independent,' I told the MD.

'You'll never last, Peter. You don't have it in you. It's a very cold world out there!'

'Excellent!' I had no desire to prolong the conversation. 'Ever since I was a teenager, people have been telling me I'm going to roast in hell. A bit of cold will be a nice change.'

Without waiting for a reply, I left and headed into the unknown.

Darth and Me

'Dr Aylett!'

'Dr Scott-Morgan!'

We'd only ever emailed before, but this seemed like as good a start as any to our video Skype:

'There is no one in the world I would rather be talking to about speech synthesis than you . . .'

As I listened to myself, it did sound a trifle over the top, but it happened to be true. I'd done my homework and studied every company on the planet offering voice cloning – basically, training a computer to sound like a particular person when it spoke. I wanted a computer to sound like *me*. And I'd concluded that my very best bet was a company based in Edinburgh. It was called CereProc (short for 'cerebral processing', which gave an insight to the thinking of its founders) and the guy I had managed to get a call with *was* one of the founders, as well as the chief scientific officer (this made him sound like a crew member on board the USS *Enterprise*, which could only be a good sign).

In my bid to change the world, I'd undertaken a trawl of the relevant research and technology development around the globe. Some (relatively accessible) was in the USA, some (more difficult for me to access) was in Japan, some (conveniently) was in Europe, some (very conveniently) was in the UK. But wherever they were, I'd realized I'd need a calling card. So I'd recorded a short video about what I was trying to do, explaining that I was trying to band together a Rebel Alliance to overturn the status quo. For the first time, I'd launched myself on social media. And I'd begun to shamelessly leverage the fact that I had a Channel 4 film crew in tow. Indeed, some of them were up in Scotland now, recording the other end of the Skype call.

'So, Matthew, tell me all about your research!'

I watched as he enthusiastically described some of the cutting-edge work they were doing. Subconsciously, I registered him as about ten years older than me; then I remembered my actual age and recalibrated him as ten to twenty years younger. But he had the energy and irreverence of a teenager, and over the next hour I grew to like him immensely. This may possibly have been slightly influenced by the fact that he agreed to try to build the best possible computer synthesis of my voice . . .

A few weeks later, Francis and I drove up a steep incline to a large recording studio on the edge of a country house estate surrounded by fields. My film crew was already in position. After the obligatory retakes of me descending on the telescopic lift of our new WAV, we entered the building and met the sound engineer. Owen was incredibly professional, but also incredibly caring.

After our first five-hour session, we'd got a lot recorded. CereProc had given us thousands of phrases to record that between them used every possible combination of sounds I was ever likely to want to make in the future. It was the same set of phrases used for commercial systems such as satnavs. In addition, though, I was to record a number of phrases of my own choosing that I might want to use in their entirety. It was one of these that delayed proceedings.

'If I recall my vernacular verbiage correctly, the apposite terminology for such an occurrence is . . .' I then proceeded to list a variety of options for the ending: 'a screwup, a shambles,' in increasing severity via 'complete bollocks,' until the ultimate, 'a cluster-fuck'.

'Sorry, I picked up some noise. Can we do that again, please?' a voice behind the camera pleaded.

'My sound was fine,' Owen countered.

'No, it was me!' Matt, my producer/director, sounded deeply apologetic. 'I started laughing.'

Pinewood Studios, on the outskirts of London, has been used for *Star Wars* and James Bond films. It was also where Channel 4 were now filming Francis and me driving down Goldfinger Avenue. We were here for my face to be future-proofed by turning it into a super-high-resolution avatar.

A wonderful person called Amanda (director of Optimize3D, one of myriad companies on the Pinewood lot) had gone out of her way to enrol a large handful of boutique firms that between them should be able to contribute the diverse skills and technology needed to create an avatar that potentially would be of the same quality as the one they'd created of Princess Leia after Carrie Fisher, who played her, had actually died. I wasn't planning to die, but my face muscles were.

There's a huge motion-capture stage at Pinewood. It's tall and well lit and incredibly empty. In most cases, everything in the final shot will all be added by computer – even the actors, who'll be turned into aliens or animals or alien animals. The cameras only record *how* people move. And in my case, it was only the camera trained on my face that mattered. It was filming dots. About thirty of them.

These hi-tech freckles had taken almost an hour to apply. A rather large guy – who I was reliably informed was one of the world's experts in applying dots to famous faces, and no doubt was usually paid a fortune for this somewhat esoteric skill – accomplished my transformation with the aid of a small bottle of special Max Factor mascara that he insisted was by far the best thing for dotting someone up. He'd consult a detailed diagram on his computer that displayed the set-up being requested, dip his brush, slowly bring his hand to my face and carefully, very carefully, dab. Thirty-odd times.

After I'd pulled every facial expression known to mankind, spoken numerous phrases prompted by a computer screen and done everything that the disembodied voice of a director somewhere requested, my carefully applied make-up was skilfully removed and I was myself removed to a photo booth in another studio.

As passport photos go, mine was the ultimate. More than fifty high-definition stills cameras wrapped around me in a globe. An expert specialist in facial expressions sidled into the booth, coached me on how to make the right expression and sidled out again. All the studio lights were turned out and after a countdown there was a brilliant shadowless flash that left me blinded for a few seconds. Then we went through the whole process again. And again. About thirty times.

I realize that the extraordinary comfort of knowing that I had at last preserved my facial movements – my personality – ought to be my abiding memory of my day at Pinewood. It certainly makes it to a strong second place. But the number one position has to go to the moment when, against all expectation, I got to cradle Darth Vader's shiny black helmet. One of the directors had it in his office. It was the actual one worn by David Prowse in *Episode V: The Empire Strikes Back*. It was enormous.

As we drove into the Imperial College campus and I caught sight of the Queen's Tower, towering as its name required,

I felt a pang of loss, the sense of an opportunity wasted, a career misspent. I'd loved academia. In another universe I'd stayed on at Imperial, never intellectually prostituted myself as a globe-trotting consultant, become a professor of robotics, gained tenure, written peer-reviewed papers, built academic credibility, been a *real* scientist.

'How does it feel to be back at your old university?'

Matt, as ever talking to me with a camera viewfinder obscuring much of his face, obviously wanted an establishing shot. I dutifully looked to the side of the camera (never into the lens, he always insisted) and told him the truth.

'I haven't been here for more than thirty years, and it feels wonderful to be back!'

There were robotics labs all over the campus these days, and I was visiting two of them. The first visit involved strapping my right arm to an assembly robot and turning it on. This is not something that should be attempted at home. Indeed, there are health and safety regulations that state it shouldn't be attempted in a factory either. But this was an academic research lab, so that was OK.

The Velcro straps were actually rather tight and the process of restraining me did evoke feelings of being prepared for punishment of some sort. Or possibly execution. I smiled at the two PhD students strapping me to the very strong-looking articulated device, no doubt subconsciously hoping this would curry their favour. They smiled back. And then further restrained me by wrapping a gauntlet around my hand.

'We've never done this with anyone other than ourselves before,' one of them commented.

'It ought to be completely safe,' the other reassuringly added.

The experiment was actually rather clever. A camera

tracked my eyes as I looked at the table in front of me. I was supposed to look at a randomly placed orange; the computer would work out where I was staring, move my arm by moving the robot arm I was strapped to and close my fingers over the orange by activating the gauntlet. I'd then look at a randomly placed bowl, the robot would move my hand over it and the gauntlet would make my fingers release the orange. What could possibly go wrong?

Well, for a start, my arm could be wrenched out of its socket. This thought was top of my mind as I wondered if my assailants' arms were slightly longer than mine. I was promised that the robot would cut out in this scenario. But one of the students made some adjustments to the computer program nevertheless. Against all the odds, an hour and at least a dozen attempts later, the experiment worked perfectly. At least, the orange ended up in the bowl. We all felt triumphant, though I was probably the only one to also feel a very sore arm.

Despite my clinical trauma, I wanted to try one more experiment. We reprogrammed the arm, and this time it reached my arm out to grip Francis's hand. It was the first time in three months I'd been able to do it. It was at that moment of contact that I realized how *really* crucial it is to be able to reach out and touch the ones you love.

The second lab visit was my chance to try a robotic wheelchair using much the same technology as a driverless car. Matt was enthusiastic that this would make great television, but perhaps we should have been forewarned that this might not be the case by the casual comment that this was of course a research vehicle and I was using it at my own risk.

'We've never had anyone disabled in it before,' the professor enthused.

'It usually behaves itself,' his PhD student confided.

It worked very well, in a hugely jerky and occasionally forgetful way reminiscent of an old granny with a tremor. It even successfully circumnavigated around a screen obscuring its final destination. It was doing so well. I could imagine Matt behind his camera getting excited that this was his money shot. And then the onboard computer got confused. And the problem with it getting confused was that it went into search mode. And the problem with search mode was that it involved the chair revolving in a circle. Fast. With me in it.

The good news was that, highly appropriately, my death ride was fitted with a kill switch to stop everything in its tracks. The bad news was that said switch was on the right-hand side of the chair, and these days my right hand was paralysed.

Seeing my predicament (though necessarily in something of a blur), my friendly PhD student attempted to hit the button for me. But by now I really was revolving at speed. He missed. Regrettably, the arm of the chair did not miss him. Shocked by this physical reprimand for his programming, it took another several revolutions for him to pluck up courage to try again. With hindsight, this was probably not helped by the increasingly antagonistic instructions from his professor.

The student stabbed at the button and missed again. The chair stabbed at him and scored again. For me, the room span around again. This was turning into the ultimate battle between man and machine. And I was caught in the middle. Revolving. Rapidly. In an allegorical *danse macabre*, my loyal PhD student stood poised like a matador attempting a *coup de grâce*. And failing. And jumping out of the way. Then trying again. And failing. To the progressively voluble encouragement of his professor. This continued for what felt like a few years.

Eventually, someone, somewhere, somehow finally managed to cut the power. Or maybe the batteries ran out. Either way, the chair jolted to a stop. The room, however, continued to spin and it was several seconds before I could trust myself to speak without throwing up. Everyone was asking how I was. I assessed the situation. In a form of cosmic balance, the poor PhD student appeared to be nauseatingly embarrassed, exactly mirroring me being embarrassingly nauseated. His professor did not appear best pleased and was waiting to see which way the scales tipped.

'Well, *that* was fun! Very encouraging research. Great job!'

Speeding Up

'I've had an epiphany,' I confided. Francis and I were making our way along the seafront of Funchal. We'd been the first off the ship, so the Madeiran sun felt warm rather than hot, and I was pushing CHARLIE's joystick far enough forward to not just keep up with Francis's relatively brisk pace but also ensure a refreshing breeze across my face.

'You can get an ointment for that . . .'

Funchal had a broad promenade paved with tiny tiles. It looked beautiful. I remembered that it was great to walk on; in a wheelchair at speed, it felt like sitting on a machine gun firing at max. As a result, I sounded like a cross between a Dalek and Stephen Hawking.

'I've realized that it would have been a huge mistake for me to stay in academia.'

'Obviously!'

'Yes, but until recently part of me always wished that I hadn't had to rebel and choose Imperial, and instead could have enjoyed the dreaming spires of Oxford, revising on a punt, becoming a don, living in relative obscurity in an intellectual ivory tower with nothing to worry about but pushing back the frontiers of knowledge.'

'That's the Wimbledon still in you! You've never forgotten what it felt like to belong, to be at the centre of the Establishment. I never had that. But there's a bit of you that's always wished you hadn't had to walk away.'

'I *ran* away! Anyway, I don't.'

'Don't what?'

'Don't wish I'd gone to Oxford or stayed in academia. Well, not now. That's what I'm saying.'

'You wouldn't have found me, for a start!'

We stopped at a busy crossing and waited for the lights to change.

'I would *always* have found you! No, my point is that the last few months have made me realize how slow and narrow the academic research is in my areas of interest. Each project is essentially one or two PhD students working over three years. They're often hugely creative but they're also hugely inexperienced. And underfunded. There's absolutely no comparison with what I know goes on in the research labs of the global megacorporations.'

Across the road, an illuminated sign of a green man started beeping and counting down. I tentatively negotiated the steep slope down from the promenade and we crossed to the other side just in time to reach safety before the traffic revved past behind us.

'Like when you teleported to those labs?'

With Matt and a film crew over in Boston, USA, I'd joined them at the Draper Laboratory from the comfort of my own home (with a second crew filming in Torquay) thanks to a telepresence robot that, via a broadband connection, allowed me to see and hear everything as if I were in one of the labs, let me look around and move wherever I wanted. I'd even held a conversation while walking down a long corridor. It had all felt very natural.

'They were amazing!' I'd known of the Draper Laboratory since I'd learned as a kid that its scientists had devised the navigation computer that safely got men to the moon and back. 'But even they aren't doing the research that's needed to change what it means to be human. Somehow, I need to tap into the best of the megacorporations. If we're going to make a difference, make a breakthrough, everything needs to be *far* faster and more ambitious than anything I can find in academia.'

We started up a steep side road, congested by café tables, pedestrians and the occasional delivery truck.

'But why would any of them get involved? Above all else, they're commercial organizations. What's in it for them?'

'I don't know. But I do know that I've got to up my game. Somehow, I've got to get the message out in a way that excites some of the best brains in these cutting-edge organizations to join our Rebel Alliance. I'm going to use my final public speech as a test. See how the message flies. See if anyone takes me seriously.'

'Oh, do you think that's wise? It's a medical conference. You're giving the keynote. Didn't Tracy ask you to "leave them on a high"? They're all healthcare professionals, and you know how conservative the clinical community can be. They're just not into AI and robotics and avatars. If you go too far, like you do, you'll turn them off. They could turn hostile. And it's your last ever big speech. It should be a triumph. Not a confrontation. If you make it too science-fiction-y, they might laugh. Or boo.'

'Then it'll be a good test!'

A white van had crept up behind us and was now tailgating me. The driver apparently did not consider a predominantly pedestrian precinct to be a suitable location for a wheelchair. Or maybe he just didn't like the disabled. He revved his engine and beeped his rather high-pitched horn. Francis wiggled between some tables to let the van pass. Under the circumstances, this felt unduly polite to me. I held my position. And flicked a switch and pressed a button four times.

When Permobil had first taken my spec for CHARLIE, they'd asked me how fast I wanted him to go. I'd said: as fast as possible. When he was delivered, and I'd enquired about his top speed, I'd been cautioned only ever to use it on private land — because it was double the legal speed limit. This had suited me fine.

Bullied by the beeping horn, the crowd of pedestrians blocking the van's progress partially cleared. This seemed as good an opportunity as any to put CHARLIE through his paces. After all, this *might* be a private road. Anyway, the Highway Code in Madeira was no doubt different than in the UK. And the tarmac surface looked very smooth.

I waited for the van to beep once more, and then I rammed CHARLIE's joystick as far forward as it would go. Instantly, the sophisticated proportional-control software sent maximum torque to the drive wheels. The tyres gripped the tarmac and, like a bat out of MND hell, I shot forward with the G-force of a catapult.

Steering at increasingly high speed through an open gap in the parted crowd, I was pleased to discover that even after three seconds of rapid acceleration, CHARLIE was yet to reach his maximum velocity. Above the soft purr of my electric motors, I vaguely heard the guttural roar of my assailant attempting to keep up with me.

Sadly, his diesel engine appeared ill-suited to the task. Worse, the escape route I was effortlessly navigating turned out to be too narrow for his van, and his frustrated beeping and revving rapidly faded into the distance. CHARLIE, in contrast, was only now reaching his stride and was still accelerating as we neared the top of the slope. I released the joystick, the regenerative braking kicked in and (with me secured by my safety belt and lateral supports) we executed a perfect skid-free, straight-line emergency stop. As trial runs went, this had to be considered a success.

Rainbows and Ghosts

We live our lives chasing rainbows and fleeing ghosts.

At least, that's how it feels to me. And I have a theory about how it all works. As far as I can see, everything tracks back to the Unwritten Rules of Being Human. My thinking goes something like this: yes, we all live our lives chasing rainbows and fleeing ghosts. And that's OK – we're all dreamers and we're all haunted – it's part of what makes us human. But our hopes and fears are not the most important parts of being human; how we *respond* is what defines us, and also defines what it fundamentally means to *be* human.

The things to remember about responding to ghosts and rainbows is that each of us can be scared, but each of us can also be stronger than we believe; each of us wants something impossible, but our secret dreams can be more inspiring to others than we imagine; each of us can find ourselves simply staying alive, but each of us can also choose to *thrive*.

The ghosts *will* sometimes catch us, and when they do, we should embrace them, and they will lose their power to bully and intimidate us. The rainbows *will* sometimes be obscured by a storm, and when they are, we should light a beacon, shine it into the deluge and make our own rainbows.

Above all, we should remember that when our response to the innermost clash of hope and fear is to deliberately break the rules of the status quo and push back against fate, then sometimes, impossibly, we *can* change the universe.

As islands, none of us can pull off such a prodigious feat. But, interacting with the right freethinking people, influencing a growing alliance of rebels, breaking free of the past and forging a new destiny, any one of us *can* rewrite the future.

In our corner of the galaxy at least, only humans are able to do this. Other advanced primates have grammar and can tell lies, they suffer from depression, use tools, demonstrate complex learning abilities, and they plan for the long term. But only *we* deliberately break the rules – or choose not to. It's what has made us indomitable.

Breaking the rules is the breathtaking attribute that more than any other has powered the ascent of civilization. It's a uniquely human balance of self-serving creativity matched with altruistic restraint, of audacious implementation with intuitive risk assessment. In a crucial sense, this trait defines both our species and our society: *deliberately breaking the rules* is what makes us human; agreeing which rules we choose *not* to break is what makes us civilized.

I'd been speaking for thirty minutes already, and despite the microphone and sound system, my voice was very tired. This was an alien experience for me: I'd always prided myself on being able to talk unamplified non-stop for a good many hours to a large room of people and never feel any fatigue. This time, with ten minutes to go, I'd be lucky to make it to the end without a coughing fit. At least Francis was sitting in the front row with a beaker of water, just in case.

It was a packed lecture theatre. Posters dotted around the walls indicated that this was indeed a medical conference, in case any of the 150 adults in the room had forgotten why they were there. So far, they'd been a great audience – they'd laughed when I'd hoped they would, they'd looked serious when I'd been brutally honest about my condition. As a result, I was increasingly tempted to gamble that they were worth the risk of giving them 'ending two'.

I'd memorized totally different alternatives. Indeed, I'd had to memorize the whole speech because, only two weeks earlier, my hands had become sufficiently paralysed that I couldn't even use prompt cards. Anyway, it was my last big speech. It was worth the effort. But were they up for the full rebel cry? I'd decide after the next bit. It was a rerun of my favourite imagery, which I'd shared with Francis when he'd asked me about my wild ideas. That was the test.

'Imagine my life in a few years' time. While I wait for a cure, I will walk again, across a grassy mountain plateau to the very edge of an enormous precipice, exotic birds calling in an indigo sky, and there I'll stand hand in hand with Francis, forever young, looking out over a beautiful alien landscape in a galaxy far, far away. And then we will effortlessly fly to the next peak to watch twin suns rising out of a turquoise ocean with an impossibly perfect sunrise. And in those moments . . .' I paused for a beat: this was the test, '. . . we will be free.'

They were doing more now than just listening intently. I was well aware that, no longer able to wave my skinny arms around, I was nevertheless attempting to speak with passion. But my voice was weak and breathless like that of an old man. Despite that, the audience was hanging on my every word. A few had their mouths open; many had tears in their eyes; several were smiling; some were smiling *and* had tears in their eyes. They had passed the test. They were ready for 'ending two'. I eased into it.

'My future's really not going to be all that bad . . .'

Time to signal the start of my peroration.

'Well, with that image, it's time for me to draw to a close, not least because, as you can probably hear, my breathing has taken on a rather heart-rending pant.'

It had.

'So, let me end with my most important glimpse of the future . . . You know, I gave my first ever major speech as a PhD student way back in 1983, in front of a thousand delegates at a robotics symposium in Chicago. Back then, I was incredibly optimistic. I can still remember the sense that a really exciting future was beckoning me on. I was convinced that if only we were smart enough, and brave enough, and got to play with enough awesome technology, then, whatever fate threw at us, we could rewrite the future and change the world. I was standing tall.'

I remembered it as if it were last year. Much larger audience. Another universe.

'Well, it's now thirty-five years later and I'm about to end my *last* ever public speech – at least the last speech where I do any work. In a Hollywood blockbuster, this would be a *poignant* moment. After all, the script practically writes itself: poor "unstoppable" Peter, finally brought to his knees. A victim, about to be robbed of everything he's just demonstrated, never to use his voice again, never to convey emotion or personality, never to reach out and touch those he loves, *never* to stand tall again.'

This was the first time in my whole speech that I'd got real. The audience looked as if I'd slapped them in the face.

Good.

'After all, that's the imagery we've all been *conditioned* to accept. Unquestioningly. For every generation before us, diagnosis with motor neurone disease has led only to a very *dark* place.'

There was dead silence. I let it dwell as I silently counted five seconds. It was painful.

'But . . .'

Again, I let it hang. Then, in as loud a voice as I could muster, laughingly, joyously:

'*But* we are a *new* generation. This is a new *dawn*. And I *still* sense a really exciting future beckoning me on. I am *still* convinced that if only we are smart enough, and brave enough, and use enough awesome technology, then, *whatever* fate throws at us . . .'

I paused, so they would have time to make the link with my own predicament.

'. . . we can *still* rewrite the future and change the world.'

I really hadn't wanted to do this next bit in front of Francis, but I'd given him the gist and he'd said to go for it. The audience needed to hear the unsaccharined truth.

'Yes, within a year I'll likely need a tracheostomy, and on that day I will speak my very last words. And yes, within a few years I'll likely be fully paralysed yet my brain fully active. And yes, of course, this is *not* a future I'd have chosen, especially not for my darling husband.'

Some of the more empathic members of the audience were starting to find this excruciating. But I needed them *all* to feel the pain.

'But I'm not going to look away and pretend it's not coming. I'm not going to look back, yearning for what I can never do again. And I'm certainly not going to *fear* it! After all, think about it: like every bully, MND's power to terrorize depends on intimidation – for MND, it's the tired old threat that if I dare to stay alive then, for as long as I can stand the torture, I'll be permanently trapped in that ultimate straitjacket . . .'

The next four words were like a very slow drumbeat marching someone to execution.

'. . . my own living corpse.'

I had to show them the instruments of torture before I

offered them a reprieve. I had to let them suffer for a few seconds more. Then I destroyed my own absurd creation.

'*Really?* In the twenty-first century? With what I know of hi-tech? I'm sorry, we're not in a Gothic horror movie.'

In a burst of relief, they realized they'd been conned. They laughed for the first time since I'd started to wrap up. A deep laugh. A knowing laugh.

'This time, MND chose the wrong guy.'

A longer and louder laugh. Someone even cheered.

'I don't even *like* bullies.'

The laughter had been running on and I'd talked over it. Now it erupted. After a short pause, I broke into it again.

'I'm sure as heck not going to respond to out-of-date, melodramatic, almost medieval scare tactics that rely on nothing but a primal fear of being trapped and helpless.'

It was quiet again, but everyone was smiling as I continued in a tone that suggested I was explaining how I'd steal the Crown Jewels:

'I won't struggle as I'm trapped – but as I'm forced into my straitjacket, I'm going to smuggle in more hi-tech defences than the White House. And then I'm going to keep adding to them. You're looking at the prototype of Peter 2.0 – but I've got more upgrades planned even than Microsoft.'

More laughter.

'I'm not dying, I'm *transforming*.'

More cheering.

'Things will get much worse for a short while. But soon after my tracheostomy, my quality of life will steadily *improve*. Then *accelerate*.'

They were ready for the finale.

'This is terminal disease like you've never seen it before. As far as I'm concerned, bring it on! It hasn't even begun to bring me to my knees. Let me now show you why, even long after I'm locked in, I will *still* be standing tall . . .'

I stopped talking and looked triumphantly around the auditorium. Then my voice boomed out again, filling the room. My voice – albeit a bit synthetic. Clear. Strong. Young again. But my lips remained tightly shut. And CHARLIE began to slowly unfold with me attached, like someone rising from a seat.

'It is simply because, thanks to hi-tech – I *will* talk again. I *will* convey emotion and personality. And I'll reach out and touch the people I love. And you know what, I will not be the only one. Over time, more and more will choose to stand beside me.'

A growing number of the audience began to realize that CHARLIE was rising to a fully standing position. Their smiles broadened.

'And we will all stand tall. And we will stand proud. And we will stand unbowed. And we will keep standing, year after year after year after year after year, because we *refuse* simply to "stay alive".'

By now CHARLIE was fully upright. I pushed the joystick to glide forward until I was towering over the front row.

'We choose to *THRIVE* !'

And, to my surprise, I experienced my very first standing ovation.

Peter's Third Rule of the Universe

LOVE
– eventually –
CONQUERS ALL

All Time High

Thirty years and ten months after the schoolboy rebellion in which I spat myself out of the Establishment and into what I had assumed was permanent exile, I found myself invited in from the cold. And one of the implications made me really nervous. After all, I'd never asked anyone to marry me before.

In nine months' time, the law was going to change: For the first time in British history, two men or two women would be able to register as civil partners with identical rights to those of a married couple. It was a *huge* milestone. Apart from the obvious symbolism, there were the practical issues of spousal rights on pensions, access rights in hospital and the ability to transfer money or property between each other without being taxed for the privilege.

What's more, I sensed that it was the thin end of the wedge and would eventually lead to unequivocally equal marriage – with no differences even in the name. And then the interminable battle would finally be over. I wanted Francis and me to ride that growing tsunami. I was about to ask him. I was also becoming incredibly nervous.

What if he said no?

It was the twenty-sixth anniversary of the day on which my universe changed. We'd celebrated our twenty-fifth with a long voyage up the Nile, visiting every ancient Egyptian site we could. But this time we were just relaxing in our cliff-top home on the outskirts of Torquay. I'd just got back from a trip around Lake Lugano in Italy with my events team (far too cold, Francis had said, for him to accompany us), scouting out locations for a private gathering of CEOs later in the year. The previous autumn, I'd hosted them in Venice, where we'd travelled in a convoy of ten gondolas (which ended up halting all water traffic around the Santa Maria della Salute) to dine in a palazzo overlooking the Grand Canal, sung to by a modern-day 'castrato'. Somehow, I needed to top that. But far more importantly, I needed Francis to agree to marry me.

I put my champagne flute down and without fuss got up from the dining table. As the track that was softly playing ended, I started a preselected tune. It was the slow instrumental version of 'All Time High' – originally a James Bond theme, which included the lyrics 'we'll change all that's gone before, doing so much more than falling in love . . . we'll take on the world and win' – right now, for us, the most romantic and personal theme tune we could imagine. It was my best shot. I reached into my pocket and took out the gold ring, just like the one fashioned from Avolean gold that Rahylan had given to Avalon when he asked him to marry him. The inside of the band was engraved with 'All Time High'.

I knew exactly what to do next. I'd imagined the scene repeatedly throughout my teens. Rahylan knelt down on one knee in front of Avalon and said:

'Will you marry me?'

Francis said nothing. For ages. Or maybe it just felt like ages. Then he looked a bit confused. Then he saw the ring. Then he got tears in his eyes. Then he said:

'Yes!' He almost couldn't speak. 'Please!'

We both took for granted that it would be a low-key event. By the time it took place we would have been a couple for almost twenty-seven years so it was hardly going to be very newsworthy to any of our friends or family. In addition, the date the law was going to change was 21 December 2005 (and, we agreed, it would be nice for our wedding to take place on the first day it was allowed, if there was any room left), so the weather would be terrible and everybody would be caught up in pre-Christmas madness. We'd pop down to the register office with a single witness, sign the appropriate form and be home for a cup of tea minutes later.

Then, on local TV, a dean made disparaging comments about civil partnerships. Francis, watching the evening news, made equally disparaging comments about the dean. Yet even at this point, our plans for a low-key event might well have remained intact had not other men of the cloth, on national TV, made vehement attacks on the very idea of gay relationships being considered comparable to straight relationships – let alone that such abominations be recognized in law.

'The Church is now the only remnant of the Establishment left fighting us!' Francis explained to our friend Vinny, using the same wording as he'd used to me the day before.

'Are you sure you're not misinterpreting their meaning? Look at me with that girl; I was convinced she wanted me.'

'You're lucky she didn't get an injunction against you!'

'That's unfair! Every time I met her, she kept smiling at me and looking at me and asking about me. It was very easy to misinterpret her intentions.'

'You were in her surgery. She was a dentist.'

'Even so . . .'

At this stage, we'd known Vinny for several years, during which time he'd increasingly become a bit of a stray cat. He'd turn up, invite himself in for a cup of tea, and more often than not fall asleep on the sofa. Then, when he was ready, he'd get up and leave. He was an island unto himself. The greatest insights into his character had been when he'd revealed that secretly he'd like to be a robot and his dream was to live on his own in the middle of Dartmoor. His favourite music sounded like a hyena being eviscerated.

'The point is, Vinny,' I said, 'the Church has had a couple of millennia's practice in persecuting gays, so it was never going to give up now that it's got so good at it. And it's even more dangerous now that it unexpectedly feels cornered and that it's suddenly losing power over people.'

I'd been monitoring what had been going on over the last several months as the change in the law became more and more likely. I hadn't liked what I'd seen. It felt like the Church was attempting to bully the State. I tried to explain what seemed to be happening.

'The Church is being strategic. It wants to cause an outcry *now* over civil partnership because it's absolutely desperate to stop the possibility of equal marriage in its tracks. That's why some of them are now going to get nasty and personal about people like Francis and me. We've unintentionally become a symbol. If society ever lets gays like us legally use the term "marriage" – just like any straight couple – then we abominations will have won. We'll be *part* of the Establishment. And all those vile religious bigots will for the first time ever be on the *outside*. So they've *got* to stop us.'

'Did I tell you my bruise is still there?'

'What?'

'It's going black! I wonder if I should go to the doctor?'

'You tripped, three days ago. That's what bruises do!'

'Even so . . .'

After Vinny had gone, we took stock of the situation. Francis had never been one to seek the spotlight, and this occasion was no different. However, stronger than his desire

226

for privacy was his refusal to be intimidated. And now, our decades-old mantra The Scott-Morgans Don't Give In To Bullies took hold. He was in fighting mood.

'This isn't just about us any more. This is about all those couples who aren't as lucky as us, who don't feel as strong as us, who don't feel as able to stand up to bullies. We need to be a symbol. We need to show that there's hope. We need to make a noise!'

The potential volume of that noise increased dramatically when we discovered that we'd been chosen to have the very first slot on the very first day. And then the wonderful people at our local Register Office bent the rules and offered to hold our ceremony in a huge ballroom in a mansion modelled on Versailles. And that meant we needed at least a hundred guests to do the room justice. And, though Anthony couldn't make it because he was just starting a top job in America running an opera house (which was a shame but no great surprise as he'd always said he would), more and more friends and family said they'd be there.

And then the Elected Mayor of Torquay offered to be one of the witnesses. And then the Ceremonial Mayor did the same. And said he'd bring the Lady Mayoress. Both in full chains of office. And then the Chair of Torbay Council asked if she could be a witness too. And the Chief Registrar said OK, under the circumstances he'd change the form to allow three witnesses rather than the specified two. And then the local press got involved. Then BBC Radio. Then ITV News. Then BBC Television News.

I felt the need to type up a basic running order for our increasingly complex ceremony. Ten single-spaced pages later, I was getting close. But there was still the unresolved issue of how Francis and I would get into the room.

'I am *not* walking down the aisle in front of all those people!'

He had been quite consistent on this for the past two months. He remained adamant for another month. Then, while selecting tracks for a playlist to use at the reception, I played him Michael Ball singing 'Love Changes Everything'.

'Now *that* is something I could walk up the aisle to with you by my side!'

The formal invitations had clearly said 'No presents – only your presence', but the night before our wedding a group of friends treated us to a lovely dinner at the Grand Hotel, where Agatha Christie honeymooned. We were then invited to step out on to the terrace overlooking Tor Bay and count down from ten; on 'Zero', the heavens erupted into a full-scale firework display that lit up the whole bay and continued for ten minutes, burning a bright memory forever.

The next morning, more than seventy friends from around the world were seated by 8 a.m. And so were over thirty members of our extended family, covering four generations and spanning almost eighty years.

Apart from my parents, absolutely no one from my *original* extended family was in the Grand Ballroom of Oldway Mansion. But to be fair, none of them had been invited. My mother had, over the decades, morphed from Devout Christian to Partial Christian to Humanist to Partial Atheist to Devout Atheist, and held very different views from those she had expressed at the start of our relationship:

'Da and I are absolutely bursting with pride at what you two have achieved! We never thought we'd see it in our lifetime. We are *so* proud of you both! It's *marvellous*!'

And so, with cameras trained on us and Michael Ball singing his heart out, Francis and I (in formal frock coats) made our way down the long aisle in stately procession with our ushers (three of our nephews and a great-nephew, all in morning dress), a very young great-nephew, also in morning dress and bearing our rings on a cushion, and two great-nieces as flower girls in chiffon. So far, so according to my running order. And then things went off-plan.

All our guests stood and started cheering. But far more symbolically, so did those on the stage in front of us. The two Mayors and the Mayoress and the Chair of the Council and the Chief Registrar.

The Establishment.

That image was still burned into my mind minutes later as Francis and I stood on stage and the official part of the ceremony began.

'Today, Francis and Peter will affirm their love and publicly declare their commitment to each other. And today, for the first time ever, it will be possible for me as Superintendent Registrar to declare that union as recognized and protected under British law.

'There is a wonderful symbolism about this historic day falling on 21 December, for today is the day of the winter solstice – the shortest day of the year. Every day that follows this one is brighter. From time immemorial, this day of the year has symbolized the end of the old and the beginning of the new. And so it is today . . .'

Francis and I formally exchanged pledges and were handed our rings, and the three official witnesses signed what they were supposed to sign. And then the ceremony reached its climax. The Chief Registrar continued:

'In a moment, you will formally be recognized as partners in law. Normally, I might just say the words. But this is no ordinary day, and yours has been no ordinary journey to get here. So, in recognition of that, I will let you choose the instant at which your new life together symbolically begins.

'You will signal the ending of your former life together, in just the same way as you began it so many years ago – with a kiss. And I declare that it is at *that* very moment that we will all formally recognize you as partners in law.

'Francis Scott-Morgan and Peter Scott-Morgan, your experience of living day by day as legal partners is about to begin. Go and meet it gladly – for your long voyage towards formal recognition is now over. With the full blessing of the law, and the total support of all here today, please now take the very last step needed to complete your journey.'

And just as I'd always imagined, in front of the whole court, Rahylan and Avalon kissed.

'Sealed with a kiss' was the headline of the TV news coverage; 'First gay wedding' was the caption over our picture. And everything that the media said about us was supportive and kind. The Establishment really had changed. But perhaps the most insightful piece was by a journalist called Ginny Ware (for the *Herald Express*, a local paper), who dropped into our lives for one morning and then disappeared again, leaving a gem behind her:

> Watching Francis and Peter Scott-Morgan achieving the same status, rights and legal acceptance as married heterosexual couples was a privilege.
>
> They have spent the past year planning their special day, which was originally to be a low-key affair with just a handful of guests and no press coverage. For Francis, who is more reserved than Peter, the decision to go public was not an easy one. It was the death of a gay friend from AIDS, which left his partner with no financial rights, that led him

to take the brave decision to be loud and proud about his union with Peter.

From that day, the pair embraced the chance to share with a sometimes suspicious and disapproving outside world the end of their long struggle to be recognized as a loving couple.

By welcoming the prying eyes of the press and TV cameras, they grasped homophobic prejudice by the throat and throttled it.

As Peter said: 'What matters is not the race, religion or the gender, what matters is the love.' The pair spoke their vows with confidence and dignity and unfalteringly acknowledged their love for all to hear.

Family and friends travelled over land and overseas to support them, even though the ceremony took place at a totally uncivilized hour on a school-day morning.

But I'm sure the guests would say they were honoured to mirror the commitment Peter and Francis have shown each other for the past twenty-seven years.

Together, they have cherished, nurtured and respected their relationship and achieved something all couples strive for – a true and steady love and friendship that has stood the test of time.

And they are right, nothing else matters.

Matt Moment

Matt and I had a running joke.

'We both *know* that for a Perfect Matt Moment you'd like me to cry at this point!'

'I absolutely do *not* want you to cry!' He'd look suitably sheepish and pull the little half-grin we'd got so used to over the six months he'd been our producer/director. 'It's just that it *would* make great television.'

'Tough! You'll just have to use your superpower to generate an *ordinary* Matt Moment.'

This was the informal taxonomy we had co-created over our half-year of accelerated friendship: while the elusive Perfect Matt Moment would require multiple poignant tears to roll down at least *my* cheeks but ideally Francis's as well (if we were looking especially crushed and defeated at the time, that would be an added bonus), the far more common Matt Moment was often triggered simply by Matt pointing his camera.

This had all become apparent a few months earlier. By that stage, my arms had largely shut down, my lung capacity was less than half what it had been at the start of the filming and my head was starting to loll. The NHS had generously concluded that I was eligible for continuing health care, and I therefore now had a couple of personal assistants who spent three hours getting me up in the morning and two hours putting me to bed. Matt, typically, asked if he could film the whole of both rituals. Twice. For a few seconds in the final film.

'I just don't know yet *which* seconds they'll be.'

'You do realize that for much of the proceedings I'm completely naked?'

'Oh, that's OK. I'll use a tight shot. And we can always pixelate you. Anyway, it's going to go out after the watershed . . .'

He arrived at 5.30 a.m. and used the key we'd lent him to let himself in, just so he could creep upstairs in order to film our alarm clock going off. And then he just kept silently filming. Until, after about ninety minutes, he piped up: 'Can you just redo that? I'm going to change lens,' as if that alone was justification for my two PAs to laboriously extricate me from the long-sleeved T-shirt they'd only just managed to get me into.

It was soon after this that Matt's name became immortalized. Within seconds of him commencing his close-up, my T-shirt became stuck. Knowing they were being filmed, my PAs heroically persevered. Matt was surreptitiously edging in closer. He seemed increasingly interested in the proceedings, encouraging our progressively frantic antics with the words: 'Don't worry – it looks *great*!'

Forgetting this was not live TV, my PAs tried to keep up appearances by acting *very* casual. Meanwhile, as if following some primal instinct to preserve self-esteem, they both intuitively pulled harder on my T-shirt. Much harder. Somewhere between my left armpit and shoulder blade the sleeve became impossibly knotted. 'Impossibly' because there really wasn't anything for it to be knotted on. It had never become knotted before. Indeed, to this day it has only ever become knotted on one further occasion.

That was on Matt's *third* take. It was at this point that I first expressed the evident truth that Matt held special sway over the universe. Far more powerfully than Murphy's Law (or even Sod's Law, now I came to think of it), Matt armed with a camera was able to warp statistical probability sufficiently to turn a trivially simple task into entertainingly muddled Great Television. We named it a 'Matt Moment'.

Over the next couple of months, I'd discovered that Matt and his camera didn't even need to be present for a Matt Moment to occur. His presence in our lives seemed to have altered things forever. Henceforth, Matt Moments just happened. Looking forward, I suspected they always would. And every time they did, instead of feeling frustrated and irritated, I hoped I'd find myself quietly smiling as I remembered his encouraging words: 'Don't worry – it looks *great*!'

And then Matt was killed. Between Christmas and New Year. In a freak accident, he slipped and fell off a roof. He was in his early forties, and his family were utterly devastated. We were going to miss him immeasurably; the positive experience we had with the creation of this television documentary was largely down to Matt's reassuring presence and brilliant ability as a producer.

Via a conduit of cosmic coincidence, from his coma, or maybe from oblivion, Matt messaged us one final time. Within minutes of his death, our doorbell rang and a postman delivered a small parcel. It was a book about cyborgs, Matt's Christmas present to us. He'd written on the front page:

Christmas 2018
Dear Peter and Francis,

What a pleasure to get to know you both – and what a journey!
To the Future!

Love, Matt xx

I was now working seven days a week, harder and longer than ever in my career; there'd be time enough to rest when I was unable to do anything *but* rest. It didn't really help. Like a house of cards, everything began to collapse from the time of Matt's death.

It started with my voice synthesizer. CereProc and I had decided that, although the thirty hours in the studio I'd already completed were ideal for the way expressive voices would be constructed in the near future (which was why we'd chosen what was called a 'deep neural net' approach), there was a totally different process we could also use, which in the short term would give a better voice.

The only downside was that this 'unit selection' approach would necessitate another thirty hours in the recording studio, and my voice was starting to fail, and Owen at the studio broke the news that they were being forced to close because the estate was being sold to developers, and there were no available slots left before the studio shut. On top of all that, the company whose system I had intended to use to read my eye movements as I spelled out words for my voice synthesizer refused to make the small software upgrade needed for me to express emotion, using the excuse that 'there is no market for it' – which as far as I could see was a self-fulfilling prophecy.

And then there were my symptoms. Even Vinny had run out of phantom counterexamples to match my own one-way deterioration. Having touch-typed since I was eleven, I'd lost a finger a week, from my right pinky to thumb. Now I resorted to stabbing at the keyboard with a few unresponsive digits of my left hand. And I could no longer feed myself, so that honour passed to Francis. And the wonderful mouthpiece ventilator that Jon had got for me on the NHS wasn't working well any more, because of the slightly unusual way my muscles were shutting down. And Francis was feeling under increasing pressure to ensure I had enough nursing care – but he couldn't find any suitable applicants.

And then I got a cold. It was hardly even worthy of the name. Back in the day, I'd have popped a couple of paracetamol, given a major speech and only been aware of my infection because of the slightly sultry timbre added to my voice by the extra phlegm on my vocal cords. It nevertheless almost killed me. Several times.

On the first occasion, I woke from a deep sleep, already in a panic, unable to draw breath. I was wearing a non-invasive ventilation mask that usually helped pump air in and out of my lungs at night. But now it was fighting me. It felt like my two earlier epiglottis episodes but, unlike before, this time I often couldn't even steal a little air between coughs because the vent pump was on the suck part of its cycle. Instead of the whoop of stridor, I heard the tortured gurgle of someone being strangled. And my hands were too paralysed to rip off the mask, let alone grab the water beaker. I felt myself beginning to black out.

Francis was awake, heard me struggling and saved me. Just as he did on more than twenty subsequent occasions over the next fortnight. It was not fun for either of us. But it did start me thinking. In the UK, only 1 per cent of people with MND have a tracheostomy to extend their lives. But the most common reason *they* then die is aspiration pneumonia – in other words, saliva or food gets into the windpipe leading to potentially fatal pneumonia. So, I reasoned, why not completely separate my windpipe from the back of my mouth so that no saliva or food could *ever* reach my lungs? It was a major operation called a full laryngectomy, typically used on people with throat cancer. It would involve taking out my larynx (my voice box), so I'd never be able to speak without a synthesizer again. But, if my logic was correct, it would considerably increase my chances of a long life as a cyborg. My challenge would be to convince a surgeon to operate on a *healthy* larynx.

Having this thought was the first time I'd felt able to push back against fate since Matt's death. And sure enough, around our fortieth anniversary, as I took my first day off for three months and we spent a wonderful day with friends and family, the house of cards stopped collapsing and even started to rebuild. Our nephew Andrew, living next door, decided to give up his job at the airport looking after VIPs and take up the full-time challenge of looking after the decidedly non-VIP Peter 2.0 instead. It turned out he was brilliant at it, and he was soon an indispensable prosthesis for my failing limbs.

'I'm thinking of writing a book about how all this came about,' I told him soon after he started, having got wholehearted support from Francis for the idea.

'Will I be in it?'

'Only if you type what I dictate . . .'

The problem with an autobiography is that, unless you're already a celebrity, no one is going to print it unless a good agent can persuade a publisher to do so. I concluded that I needed an agent. Of course, the problem with an agent is that, unless you're already a celebrity, no one is interested in representing you unless a publisher is going to print your book.

There was a similar catch-22 when it came to persuading megacorporations to get involved with my ideas. Very reasonably, they wanted to work with philanthropic organizations rather than individuals, but none of the philanthropic organizations that I knew had any interest in my ideas – let alone in working with megacorporations.

But here I might have had a breakthrough. In my role as a trustee of the MND Association, I'd been banging on about thriving with hi-tech. If only to shut me up, I'd been invited to chair an advisory group on the topic. I'd suggested we instead create a think tank and try to interest a few elite IT corporations in my ideas. We'd visited a manufacturing firm and I'd excited their CIO. With his blessing, a guy called Ray in their Innovation department offered to put me in front of some of his contacts in IT. It was then up to me to inspire them with my proposed streams of research.

I was elated when I reported back to Francis.

'Be careful,' he warned. 'If something seems too good to be true . . .'

'. . . it often is!' I chorused with him.

Meanwhile, in a remarkable twist of good fortune, Owen the sound engineer turned out to live a couple of roads away. His cat often strolled through our garden. Based on this proximity, Owen set up a temporary sound studio in our lounge and our recording programme resumed. Our luck continued when another near-neighbour, Nigel, agreed to join my team of PAs and help take up the growing strain of my care needs. Things were finally looking good again. In celebration, I took the whole of my sixty-first birthday off. Again, we spent a glorious day with family and friends. Vinny couldn't make it, for the second year running.

'The selfish sod could at least have texted you to say happy birthday,' Francis commented the next day.

'I know, but what if this time there really *is* something wrong with him?'

'Trust me, if Vinny even pricked his finger, we would be the first to hear all about it!'

Then things *really* got going. Based on some sample chapters and against all reasonable expectations, I was signed by a top agent, Rosemary. And then the top ENT surgeon at NHS Torbay – a wonderful consultant called Philip – agreed to see me, agreed with my logic and agreed to give me an elective laryngectomy, and, in a meeting with Francis, Andrew, me, Maree (my wonderful anaesthetist), and Jon (my wonderful respiration expert), agreed a date – as late as possible (so I could try to arrange some hi-tech to take over my speaking) but well before the winter (so I could try to avoid any more colds). We all checked our diaries and chose Thursday, 10 October 2019: five and a half months away.

'I don't want my last ever biological words to be "ten – nine – eight . . ." as I drift into anaesthesia,' I explained to Maree.

'No problem. You can nod or shake your head to any questions after you've said what you want to say,' she reassured me in her usual Anything Is Possible voice. 'Do you know what your last words are going to be yet?' This was in far more of a schoolgirl-gossip tone.

'I've known for at least a year.'

Perfect Matt Moment

'What do you *mean* he doesn't want to see us again?'

'That's what he said. I kept texting him, asking if he was OK, and eventually after three weeks he got back saying he found it difficult being around you.'

'But he always said we were his best friends!'

'Yes, and we thought he was one of our best friends. Obviously not.'

Francis sounded as dejected as I suddenly was. We'd known Vinny for decades. We'd often described him to people as totally trustworthy. Loyal.

'Did you text him back?'

'Yes, I told him that if he was so cruel and selfish then he could fuck off!'

'And what did he say to that?'

'He never got back.'

For some reason, this felt more of a body blow than my diagnosis had been. Over the next few days, we both felt like the stuffing had been knocked out of us. Betrayed. Abandoned. And this devastation seemed to trigger a tsunami of setbacks and aftershocks.

I was increasingly realizing that the upgrades planned for CHARLIE wouldn't fit with what I now hoped some of the companies Ray had introduced me to might offer to do. But that was the least of my gnawing concerns. Or maybe I was just slipping into paranoia. After all, these days I was getting a couple of hours less sleep each night than had sustained me all my adult life. Either way, something strange was happening on the work front. Now that I had stirred up interest in my research, Ray seemed to be cutting me out of the loop, having side conversations with the MNDA about my ideas, positioning himself as the gatekeeper to the other companies. Then someone briefed the media that Ray was the lone person responsible for everything the think tank was contemplating. Everyone was still being immensely friendly to me. But I couldn't help feeling suspicious.

Then I heard that people were talking about ways to make money from the think tank's research – my research. Don't get me wrong, I'm not opposed to anyone making money. But in this case, the more I thought about it, the more I was convinced that everything we did should be open source, available to all. What I was hearing disappointed me and made me uncomfortable.

On the home front, Nigel (the guy who had agreed to join our team) sent a text saying he'd changed his mind. Francis would now have the added pressure of trying to find a replacement. The news came on a bad day. We'd just been discussing the practicality of taking what might be our final cruise to the Caribbean, long booked for January 2020. With my laryngectomy now scheduled for October 2019, we very reluctantly conceded that we had no sensible option other than to cancel and lose our deposit. I knew that Francis had been holding out for that holiday. I felt unbelievably guilty. Then one of my care team left a message that they wouldn't be available that evening to help put me to bed. Again, that would fall on Francis.

'You'd better see this . . .' Andrew had been checking my emails now that I could no longer even use the trackpad on my laptop. 'It's *totally* out of order.'

It was written confirmation to every company that I'd

spoken with that the MND Association think tank was pulling out of all involvement in my research other than a potentially limited pursuit of speech synthesis.

Andrew was livid on my behalf. 'What they don't fucking mention is that a few weeks ago the bloody board of trustees excluded you from the meeting "so that people can talk freely", and with no one in the room who actually knew what the fuck they were talking about they got cold feet and stopped almost everything!'

'That's their job. They need to be seen to protect the needs of a huge membership.'

'Yeah, but some of the board are complete wankers! And what about Ray? He's gone very quiet. He was very friendly while you were useful to him but he hasn't answered any of your emails for weeks. And what about that prick who never smiled at you in any of the meetings?'

'He actually never smiled at me at any time in the last year. I assume he's homophobic. Anyway, I've resigned now. I was elected on a promise to research how to use hi-tech to thrive. If I can't do it within the Association, I'll work with the companies from outside and we'll make everything available to the members regardless. It may be easier for everyone this way.'

I was putting a positive spin on things. But I actually felt heartbroken.

And then an email finally arrived from Ray. There was none of the usual friendly beginning. It was terse. No friendly ending. No hint of regret. Simply a statement that he and his company would not work with an individual and so would not continue working with me. And he expected that every other organization connected with the think tank would take the same view. Basically, the think tank would continue, but not working on my research, and not with me. By implication, I could continue with whatever research I liked, but not with any of the companies that had shown interest so far.

Andrew had gone home by the time the email arrived, but Francis was with me.

'Can they do that?' His voice was tentative, already prepared for the obvious answer.

'They've done it.'

'But it's so unfair! You've been working non-stop to make it all happen, and you're the only one that hasn't been being paid a salary.'

Normally, in the face of perceived injustice, he would have been angry by now. Instead, he only sounded tired.

'And what about all the lost time? We could have been enjoying ourselves. We could have taken our last holiday where you were able to feed yourself. We could have tried to have a little fun, for fuck's sake. This is all miserable enough without us throwing away what little time we have on ungrateful bastards who don't deserve us sacrificing everything for them!'

Now he was sounding angrier. But it was a desperate, desolate anger.

'Fuck the lot of them! They're not worth it. We're going to take the summer off, try to relax, go places, have our last bit of fun.'

'But I've still got the book to write. I'm only just going to get it done. Even without the think-tank work, it's going to take all of the summer.'

'For fuck's sake, Peter! Don't make me resent you and your fucking MND more than I already do!'

I knew we were both stressed and chronically tired. I knew in my heart of hearts that he still loved me. But I also knew that just at that moment he didn't. And that's when I fell apart.

I couldn't remember the last time I'd broken down in front of him – maybe back in my twenties. I'd grown a lot stronger since then. More resilient, more secure. The first tear came as a surprise. The second came from the other eye and was almost as much of a shock as the first. But their unexpected arrival appeared to change all the rules. Nothing of the past mattered any more. Nobody cared anyway. Least of all me. And I just let go.

A small part of me registered that there was something pathetic about a sixty-one-year-old man wallowing in self-pity, crying his eyes out like a little child, no self-esteem, no control, no pride, no worth. A husk.

'Oh, for fuck's sake! What are you doing now?'

It was like listening to a distant echo from another dimension. It felt unreal. All that mattered was the black hole of despair, drawing me back home. I vaguely remembered it from childhood, before I gained my teenage shield of arrogance, when I was always one of the last to be picked for a team, one of the only ones not invited to a birthday party, never part of the in-crowd, always the butt of jokes. Always on the outside looking in, always lonely. I'd completely forgotten the feeling. And now, with my first tears dissolving the barriers of half a century, it all came rushing back.

I was now bawling. Short, unappealing wails, made even more alien by my total lack of breath. The last spluttering vestige of a once-proud human being.

'Now, come on, this isn't what we do!'

I tried to explain. I tried to speak. But I had no breath. I couldn't even do that. Already I was a failure at everything. And it was going to get worse. So very much worse.

'Don't be like this. We have to be strong.' The spark of anger and frustration was gone from his voice, replaced by the sound of someone starting to cry. 'We draw our strength from each other.' He started to sob. 'I can't cope if you're not strong . . .'

We hung on to each other like lone outcasts abandoned on a hostile planet. Us against the world. Forever. To the bitter end. We hugged, convulsing with synchronized sobbing, two old men crushed by adversity, no fight left, giving in at last.

We clung to each other for eternity.

'Now, we've got to pull ourselves together.' Francis was the first to successfully pull away from the black hole. 'We'll find a way.'

I tried to respond, but I had neither the breath nor the appropriate words to explain that there *was* no way left that would work in time. And if it didn't start working while I still had a biological voice, then realistically it would be too late for me to get it to work at all. I suddenly felt *so* tired. Physically. Mentally. Emotionally. It was all over; I finally acknowledged to myself that I had given up. And the weight of the failure, and the failed responsibility to all those who had believed in me, felt too heavy to bear. Above all, I had failed Francis.

'I love you,' he said, still sobbing.

'Ay —harr – ooh – ooh' was all I could manage between uncontrollable sobs.

'I know,' he said, and he started crying again. We hung on to one another, rocking back and forth, each trying to comfort the other.

Eventually, Francis forced himself to be strong for both of us.

'Now, we need to pull ourselves together! We *are* going to get through this. You and me. Like always.'

He held me very tight.

Far to the west, in Avolee, Rahylan, recently eighteen years old and so eligible to compete in the Games, had been just at the start of his overnight vigil – meditating on top of the obelisk at the summit of a huge mound that rose above the forest. He'd been engulfed in the pale-blue glow of the Flame of Analax, which forever burned from the marble obelisk, the eastern sky radiant from the recent double sunset. Though his eyes were shut, the Flame allowed him to sense everything around him, so he'd known that Avalon was coming even before the young warrior appeared over the ridge and stood at the edge of the plateau, looking at him in wonder. The Flame had revealed three possible consequences of Avalon's arrival: loneliness, death or love.

He'd sensed Avalon moving closer in his blue leather kilt and fine, loose-fitting shirt, with his shoulder-length red-gold hair, handsome face and beautiful blue eyes, until he'd been standing just out of reach on the other side of the pit that ran around the obelisk. Avalon had read the inscription on the four sides of the stone:

> Enter as one and the Flame will show
> All the futures you can know.
> Enter as two and both reveal all,
> Then leave as one or not at all.

Avalon had stood silent, staring at Rahylan's face as the sky had slowly darkened and the stars had begun to show. The Flame had seemed to burn brighter and bluer and the three moons had risen one after another, and still Avalon had stood unmoving. Shortly before the first signs of the first sunrise, Avalon had smiled and reached out his right arm until his fingers had almost touched the Flame. Instantly, one future had vanished, leaving only death or love.

Eyes still shut, Rahylan had smiled back and reached out from the Flame. Their fingers had touched. Avalon had strained forward over the pit and their two hands had clasped. The Flame had crept along Rahylan's out-stretched arm, reached Avalon and rapidly engulfed him. Then the revelation had begun: everything about each of them revealed to the other. The ultimate test. All or nothing. And everyone knew that the Flame of Analax only let you live if the outcome was all.

As the first beam of light from the first sun had streaked from the western horizon across a coral sky, Rahylan had opened his eyes for the first time and looked lovingly at Avalon. But the two futures had both remained open. Rahylan had pushed off from the obelisk, trusting Avalon to pull him across the pit to safety. The Flame had released them both as Rahylan's bare chest had thumped against Avalon, who'd immediately wrapped a protective arm around him to stop him falling back into the pit. They'd both looked deep into each other's eyes, and Avalon had spoken his very first words, slightly raspy in the cold early-morning air, leaving only one future possible:

'I'm yours forever.'

Where Tomorrow Comes From

As I lay awake in bed, in the middle of the night, my subconscious insisted on reminding me of something that felt completely irrelevant: for almost 13.8 billion years the universe was incredibly boring. Magnificent. But fundamentally mind-numbing.

I knew the sequence well; why was I wasting any time thinking about it? There was a Big Bang, then a *really* big gap, and on the outer reaches of an unremarkable spiral galaxy, about as far away from any real action as it was possible to be, a planet formed around a middle-aged star. It was an insignificant event. This was just one large lump of rock out of at least 100,000,000,000 other planets within the same galaxy. There were 170,000,000,000 *other* galaxies.

As ever, my conscious mind responded to the epic scale of the cosmos and the relative irrelevance of humanity. Lured by the glorious science involved, my brain couldn't help but complete the story. In a celestial game of billiards, something seems to have bumped into this particular rock ball and it got a moon. There were lots of volcanoes, which led to water, simple life, oxygen and, at last, complex life. From then, you might have expected an interesting future to unfold, at least on this tiny speck in space. It didn't.

Ah! *That* was where this was going! I began to understand why I had woken thinking about the universe rather than my own world crashing apart around me. Now this train of thought felt more promising. I let it run. Despite an explosion of life during which evolution really took off and intelligence slowly emerged, things continued to just happen. About 66 million years ago the dinosaurs couldn't avoid eventual extinction as a result of an asteroid crashing on to Earth, any more than humans could about 76,000 years ago when they were almost wiped out by a supervolcano.

Yes! That was the key! They were all powerless. The future just rolled over them. To this day, that pattern of inexorable inevitability punctuated by acts of utterly random unpredictability remains typical of the entire cosmos. The kindest summary of where tomorrow comes from is that nearly all of it is excruciatingly dull. Except for one occasion about 5,000 years ago. Then, one of the rarest events in the universe took place. And it happened here on Earth.

It was where our cosmic irrelevance got called into question. It was where our individual irrelevance *also* got called into question. We call it the dawn of civilization. But it's *so* much more. It's the moment that the future becomes fascinating — because, for the first time, humankind successfully rebelled against destiny. Cave dwellers could try to impact their individual lives. But they couldn't deliberately change the course of history for all humans in the way that, say, the ancient Egyptians did.

That's it! Draw on the power that that knowledge brings. Remember! Those first civilizations showed it was possible for a single individual occasionally to make such a difference that it affected everybody who followed. From the moment of that great divide in history, *it became the birthright of every human being to be able to change the universe.*

For a moment, I felt a sense of power flowing back into me. I once again dared to believe it was possible to make a difference. I'd finally worked out many of the details of how to do this by my sixtieth birthday, and I liked the whole idea just as much as I had at sixteen.

But having glowed for a moment, in the cold dark of night the embers of my self-confidence chilled again, and the whole concept rapidly returned to being one step short of gobsmackingly irrelevant at around 2 a.m., as I lay in bed, now wide awake. Nevertheless, still only a few hours since my total meltdown in front of Francis, I realized I was no longer fully wallowing in self-pity – although I certainly had several toes immersed, ready to slip back in given the slightest provocation, while my brain's executive functions were relentlessly reminding me to lose all hope.

The next ninety minutes passed faster than normal, or I unknowingly drifted in and out of sleep. Either way, by around 3.30 my brain was feeling irrationally positive; it had come up with the germ of an idea that, despite a part of me dismissing it as absurd, impractical and delusional, stubbornly insisted on continuing to germinate.

By a little before 5 a.m. the sun was up, and I was so wide awake that it was clearly ridiculous to even try to get back to sleep. Anyway, there was so much to do. My brain was by now feeling irrationally exuberant; it had given up complaining and moved on to planning.

'Are you awake?' I tentatively enquired. In my defence, Francis was definitely not snoring and he had just turned over, and for the last few weeks he'd been waking very early. In the absence of a coherent response, I bided my time. But when he made the mistake of yawning, I renewed my onslaught at a slightly enhanced volume: 'Are you awake?'

'No.'

'Oh, good! I've been awake all night.'

Francis sat up and tried to show concern despite his facial muscles not yet having got the memo to get moving.

'Are you all right?'

'Oh, yes! I'm fine. I've just been thinking all night –'

'Oh, for heaven's sake! I *told* you to leave it alone. They're not worth it. You tried. It didn't work. Move on!'

'But that's the point! I've worked out how to *make* it work! I've had an idea! Let me explain . . .'

'Not before I've had some coffee!'

Francis hauled himself out of bed. Ten minutes later he crawled up the stairs again with two perfect caffè macchiatos. Only when he'd emptied his cup did he allow the conversation to proceed.

'OK, so what's the big idea?'

'I think you and I should set up a charitable foundation, The Scott-Morgan Foundation, a purely philanthropic research organization to pursue all my ideas.'

'Well, that's obvious!'

'What?'

'If the MND Association isn't in a position to do it properly, then obviously you should. We discussed this months ago . . .'

'Yes, but it wasn't relevant then.'

'It is now. And if you don't act *very* quickly, then everything will fall apart. You've got to get it all sorted today.'

'It's Sunday! There's no one around!'

'Well, that's perfect, isn't it? I was thinking last night as I fell asleep that this way you can get your ideas in order and send lots of emails and be ahead of the curve by the time everyone comes back to work on Monday morning.'

'What?'

'I was thinking last night –'

'You didn't *know* about the Foundation last night!'

'No, but I knew you'd come up with something.'

'But what if I hadn't?'

'I'd have made you another coffee . . .'

I needed the coffee anyway, and then another in the early

afternoon. But soon after that, I'd got my ideas sufficiently in order that I could send an email to everyone I'd tried to enthuse in the preceding months, alerting them that Francis and I were in the process of forming a fully independent research foundation, and that therefore all the ideas we'd been discussing could proceed. Before hitting send, I read it to Francis.

'Aren't these the same companies that Ray said wouldn't work with anyone but him and the MND Association?'

'Yes, but that logic got blown out of the water when the Association trustees got cold feet. Anyway, the Foundation will be legally separate, run by an independent board of directors. It's not like any of the companies will be working with me as an individual.'

'I know, but will any of the companies you're emailing actually join us?'

As usual, Francis had instinctively and effortlessly found the weakest part of my plan.

'I don't know . . . maybe. I really hope so. We don't need many of the megacorporations. Even just one would be enough. If we can get one, and also attract enough of the critical smaller companies and experts, then we'll gradually attract more and more of the big corporations, and the better the Foundation does, the more the best people will want to join us.'

'So it basically all depends on whether in the next day or two you can pull a core team together to change the course of history!' Francis didn't need to say he was fully behind me, so instead he added: 'No pressure there then . . . Roughly how many directors will we need for the Foundation?'

'Eight.'

'*Exactly* eight?'

'Well, I worked out the distinct roles that we need to pull

this off. We'll need a few more when things really get going, but for the moment we need six plus us. Of those, I can take the role of Chief Scientist, and you're ideal for Director of Care –'

'Oh, no! You can find someone far better than me!'

'I absolutely cannot! You spent twelve years, for heaven's sake, *running* a *care* facility. And these days you've got as much direct insight into caring for someone with extreme disability as almost anyone. Anyway, you're the co-founder!'

'Hmm, I'll think about it.'

'So, out of our Magnificent Eight –'

'Ah, well, if we're going to be Magnificent . . .'

'Of course we're going to be Magnificent! Anyway, we've already got *The Guinea Pig* and *The Carer.* By my calculations, we need to recruit six more key characters if we're to have a shot at Breaking the Rules. Here's what I'm thinking . . .'

The Magnificent Eight

The Governor

Alun (the same Alun as had phoned during bowel prep) had been chair of the all-powerful board of trustees of the MND Association for years, but when he'd called me, he was just about to step down, having held the post for the maximum time allowed. He and I had never attended a single board meeting together. But we had kept in touch.

These days, Alun continued to chair the Merseyside branch; that was his home territory, which granted him a refreshingly down-to-earth viewpoint that I loved. He also happened to know more about the intricacies of charities governance (basically, who legally can do what) than anyone I knew. And, over the time that he'd unofficially mentored me in my trustee role, I'd grown to trust him. With skills no one else had, he'd be a huge asset to the Foundation. It being a Sunday, he made himself available for a video call within minutes.

'I wondered if you could be our Treasurer . . .' He didn't say no, so with extra confidence I added: '. . . as part of the Director of Governance role?'

He smiled.

'I'd be delighted!'

The TV Guy

The head of the BBC once gave me unrestricted confidential access to decode the Corporation's hidden inner workings. It was then that I learned that television is in fact run by a secret race of chameleons.

It turns out that anyone who makes it big as a manager has to be able to manage not only mere mortals, but also those who have made it even bigger than they have: the 'top talent' (that's 'stars' to the rest of us), some of whom have egos sufficiently massive as to exert their own gravitational field. To rise to this Herculean task in psychomechanics, the best managers have finely honed the arcane art of being all things to all egos. I am not for a moment suggesting they are two-faced; the best are at least ten-faced. Maybe more on a good day.

So when I first met Pat, more than a year before I was ringing him now, I was prepared to be simultaneously charmed and bamboozled. After all, before he'd become MD of the award-winning production company competing to film a documentary about me, he'd been head of BBC television production responsible for more than three thousand staff and an annual budget of over £400m. He probably had more faces than the Kohinoor diamond.

Then I gradually got to know him. And really like and admire him. The face he wore was remarkably consistent, but I couldn't quite categorise it. It was partly the face of a feisty campaigner standing at a street corner passionately advocating to passers-by all forms of diversity and inclusion. And equally it was the face of a privateer captaining his galleon across the seven seas in search of hidden treasure and glorious adventure. What it was *not* was the face of the Establishment.

'I wondered if you'd consider becoming our Director of Media?' Pat knew people. A lot of people. When there was a second's silence, which I worried was half a second too long, I added: 'Of course you wouldn't have to join until after the documentary was finished.' Still silence. 'To avoid any risk of perceived conflict of interest,' I concluded, as if this of all reasons might be the clincher.

'Now, Peter, knowing you and Francis as I feel I now do, can I assume that you are planning to shake things up a bit with this Foundation of yours? Even be willing to ruffle a few feathers, if that's what it takes? It's not going to be one of those boring charities, I take it.'

'I promise you, Pat, the Foundation may become many unlikely things, but boring or conservative or risk-averse will never be three of them!'

For the first time on the call I heard him chuckling. His deep bass chuckle, with the occasional tenor descant that made him sound like an overgrown schoolboy scheming mayhem.

'In that case, I wouldn't miss it for the world!'

The Voice Doctor

Since our first Skype call, Matthew and I had been in frequent email contact while CereProc built my new synthesized voice. More importantly, we'd now met a couple of times and Matthew had passed the Francis Test – an arcane mixture of intuition and deceptively insightful questioning leading to a character assessment that I'd learned over the decades tended to be worryingly accurate.

'You are such a charmer!' was Matthew's own assessment of my attempt to persuade him to become Director of Voice Synthesis. 'Sure, I'd be honoured to get involved!'

With that out of the way, we continued an earlier discussion about how to get me singing again. A few months earlier, my lung muscles and my throat muscle had conspired together to stop me singing ever again. And, having spent a lifetime singing to myself, I really missed it. A few weeks after I'd mentioned this to Matthew, he'd emailed me a file of my synthesized voice giving a fair rendition of 'Ding Dong Merrily on High'.

'It's only a proof of concept,' he'd insisted when we spoke later, 'but look up the lyrics of "Pure Imagination". They're perfect for you. There's a bit that goes: "Anything you want to, do it. Want to change the world? There's nothing to it!" Even an old cynic like me got a tear in his eye when I heard that.'

'How about us releasing a music video?'

'We'd need your hi-res avatar up and running. But yes, why not? I love the idea!'

The Avatar Geek

'So, we're going to make a music video! We just need your hi-res avatar to mouth the words to my synthesized voice. Do you think it's doable?'

One of the great things about Amanda and her colleagues at Pinewood was that there was very little, I suspected, that they considered *not* doable. And she loved working with Matthew; her very first words to me, in an introductory meeting with my friends at CereProc, had been: 'I'm in geek heaven!'

Amanda's colleague Adam had recently painstakingly added every hair on my avatar's head, every eyelash, every imperfection. He'd also taken it on himself to change my avatar's hairstyle. Francis and Andrew had agreed with him that the end result was such a great improvement on the original that I was now probably the only person on the planet to have changed his hairstyle and colour in order to match his avatar. Inevitably, this had then resulted in yet another colleague in another firm (Ari, from Embody Digital) having to change the hairstyle of my *lo-res* avatar . . .

'Yes, it's *absolutely* doable! We can get Peter 2.0 to do

anything in hi-res – it just takes too much processing for it to happen in real time. But give it a few years and we'll be able to generate hi-res Peter 2.0 in real time *all* the time!'

As it was, Amanda had already emailed her views on the Foundation before our video call: 'Of course – I would be absolutely honoured to be a trustee onboard the Rebel Alliance fleet! I truly believe the establishment of the Foundation and the collaborative forces being put together is astounding. There are just so many areas in which this consortium of great minds and headstrong, battle-ready creatives can have such a positive effect. Life changing; world changing.' I'd taken this to mean she was in geek heaven again.

The Designer

Esther came from Madrid, was based in London and spent most of her time at the moment in Paris. Wherever she was, she spoke the local language fluently. From the first occasion we met, she instilled in me total confidence that if anyone got in the way of our mission she would efficiently obliterate them without losing her near-permanent smile. We clicked from that very first meeting.

She was a director of a global design and innovation consultancy, but every time I'd met her to discuss my ideas, she'd been on vacation, or it was over the weekend, or a holiday. This had been the first indication that maybe she actually cared about what I was trying to do.

'I *really* care about what you're trying to do,' Esther confirmed, with a level of enthusiasm that only a true Latin can pull off effectively. 'I would *love*, Peter, to be on the board of your Foundation!'

This was good news. Equally good news turned out to be

that Esther's enthusiasm was clearly contagious. Soon, a couple of her young colleagues, Laura and Robin, began giving up *their* spare time to work with me to design part of the user interface to all the AI systems I would increasingly depend on. Their whole philosophy matched mine exactly – design hi-tech around people, not the other way around. The fact that they were lovely people themselves was an added bonus.

The AI Wizard

According to the laws of probability Jerry and I should never have even met, let alone become good friends. For a start, Jerry lived in St Louis, USA. I hadn't been to St Louis since that time young cowboy Brad sauntered on to the Greyhound and into my life back in 1976. Jerry wasn't even born in '76.

A further contributing factor to the improbability of us meeting was that Jerry worked for one of those companies that nobody in the real world had ever even heard of: DXC Technology. DXC was in fact a game-changing global IT consultancy. It had 150,000 employees, operated around the planet and in the previous year had turned over $25 billion. But even so, I had absolutely no link in to them.

At least, I didn't until a guy called Patrick introduced himself to Francis and me and asked the apparently irrelevant question: 'Are you interested in art?' I'd got the impression that even he wasn't convinced of the relevance of this line of interrogation, given that I'd just been talking expansively about hi-tech, cyborgs and all things scientific. I knew for a fact that nowhere on my digital footprint was there any indication that I'd spent all those teenage lunch breaks in the art building. He couldn't possibly realize the unfulfilled vacuum left by choosing the sciences over the arts. So really, it was a gloriously dumb question for him to ask. Patrick explained that DXC's Head of AI had come up with the wild idea of building an AI system that would let me create art even after I was fully paralysed. I assured him that I *loved* wild ideas.

A few days later, I had an email from the Head of AI: Jerry. Our first video call ran for over an hour. We reconvened two days later, for another hour. We then scheduled two hours a week. Every week. Indefinitely. By the time I plucked up the courage to ask Jerry if he'd consider not only joining the board of the Foundation but becoming the Vice Chair, I'd grown to love our calls together. They were fantastically stimulating.

One of us would bounce an idea to the other, who'd improve it and bounce it back, and so on. Jerry would write some new AI code based on our latest dialogue. Very quickly, we'd discovered that for me to become a genuine cyborg artist – creating art that no human *or* AI could come up with alone – we'd need to consider far broader issues ranging from how best to operate humancentric AI all the way to fundamental ethics questions. Just as exciting, we were soon experimenting with AI that no one had attempted before. It was revolutionary. It was also huge fun.

I was so concerned about damaging this dynamic that I completely wimped out of raising the topic of the Foundation during one of our calls, using the self-justification that it would divert valuable time away from our co-creation. Instead, I sent Jerry a detailed email covering everything I had in mind. I explained how I believed that, with the right people at the epicentre of the Foundation, our growing Rebel Alliance would at last be able to break the rules of the status quo and change things forever, nudging the future away from stand-alone AI towards human-AI collaboration. Ludicrous though my request might seem to him, I asked him to seriously think about it.

Less than a minute after I hit the send button, Jerry's reply pinged up:

'I thought about it (for two whole seconds). I'm in!'

Right to THRIVE

Hi, Peter,

We really urgently need your help ASAP. Julian is currently in hospital with a perforated bowel from the [feeding-tube operation], and we are coming across NHS politics which could potentially result in him not receiving the surgery that he desperately needs. We have been told that they have the right to decline level-3 treatment in ICU due to potential weaning difficulties post op. Their attitude is to sign a DNAR [do not attempt resuscitation] form and start thinking about end-of-life care. Julian is not at this stage at all; he only has NIV [non-invasive ventilation] at night and his oxygen levels are at 90 per cent without support. They are not giving us clear answers nor are they prepared to discuss trachy or lary or stoma options. We don't know what to do. If you could help in any way, we would be forever grateful as with their care and attitude he will not survive.

Seriously, who would have guessed that cheating death was a full-time job?

One of the main reasons for this, I've learned, is that the worst thing about MND is not the disease itself, it's the attitudes that surround it – the attitudes of some medical practitioners, the attitudes of some of the charities involved, the attitudes of governments, the attitudes of the general public, the attitudes of friends and families, and, most importantly, the attitudes of those of us diagnosed with 'the cruellest disease'.

I didn't really know anything about 'Julian' other than

that a year earlier he'd sent me a couple of messages saying that he found my Right to THRIVE message inspiring and reassuring. Then his sister messaged me on Facebook asking for help. It was not great timing. With less than two months before my laryngectomy, my book only half written, the Foundation still to be properly formed, and no hi-tech systems yet working that would let me talk effectively once I sacrificed my biological voice, life was rather hectic. Maybe Julian's sister had misunderstood. Maybe she was overreacting. Then again, maybe she was right. Just in case, I dictated a stream of emails for Andrew to send to everyone I could think of who might be able to help.

The reality was that, regardless of whether Julian was indeed at risk, I knew that huge numbers of 'Julians' across the UK – and indeed across the world – were dying because of the attitudes to MND. Dying not because there was no way of keeping them alive but because they were being allowed to die. Or being told to die.

I'd been immensely lucky that almost everyone within the NHS in Devon had been totally supportive of my goals and my treatment had been exemplary. But over the previous year I'd had more and more strangers around the globe contact me with horror stories of how they were being denied the life-sustaining treatment I had come to believe was something of a human right. Worse, when I'd raised this with some of the charities ostensibly protecting the rights of those with MND, I'd been told that they 'needed to be careful' because they didn't want to risk 'upsetting the clinical community'. The risk of failing to demonstrate any actual leadership appeared to be an utterly alien concept to them.

Fuelled by my bubbling frustration at the blatant injustice of it all – not least in the UK – I'd written to my Member of Parliament, Kevin Foster. He offered to meet me and lend his support. Which was fine, but we both knew that with Westminster imploding into febrile chaos over Brexit nothing would happen. And, of course, it didn't. After all, that's exactly how the Unwritten Rules of the Establishment are supposed to not work.

But that doesn't stop me seething about the way that outmoded attitudes are causing so much unnecessary pain. Some of those defeatist attitudes remain in the medical profession itself. And many charities appear to paint a relentlessly horrific picture of MND; maybe they think that's how to get more donations towards the (almost certainly distant) cure. But much of the problem comes from the resolutely negative attitude that many of us with MND find ourselves forced into because of everything we and our loved ones hear and read after we're diagnosed – especially if we make the disastrous decision to google 'MND ALS'.

Some doctors are amazingly good at communicating the diagnosis; some appear to be atrocious. I've been told of one consultant who combined the news with the verdict: 'Sometimes shit happens.' Another simply handed over both the diagnosis and a box of tissues.

Such attitudes are corrosive. As a result, many medical practitioners tell me that some of their MND patients remain in denial and don't want to talk about even having a feeding tube. That's tragic. Not least because those same patients are unlikely to be able to remain in denial as they starve to death but find it's now too late to have a feeding tube fitted.

Some people, I'm told, choose not to proceed with any preparations for when they can no longer swallow because they view having a tube attached to them as the first step towards profound disability. And they can't bring themselves to take that step. Fine. But the first step *actually* happened when they noticed their very first symptom of MND and didn't know what it was.

There is no symbolism in accepting your first life-preserving procedure – other than the heroic symbolism that you are still very much in control and pushing back against fate. People without MND have feeding tubes fitted all the time. Many of those people stop using them when they are no longer needed, whereas unless there's a medical breakthrough those of us with MND will carry on needing them. But acting as if getting a feeding tube is some sort of watershed moment risks overdramatizing its importance.

Some people, I'm told, in effect choose to die 'early' through starvation. And some, refusing ventilation, choose to die by slow suffocation (although, let me stress, these days the palliative care is so excellent that there's absolutely no need to suffer having made either of these choices). Provided that these people make their own informed decisions, then I'm perfectly comfortable with that.

What makes me desperately *uncomfortable* is the thought that maybe one or two or many or all might make a different decision if they knew everything that I do. I can never forget the message I received prior to my tripleostomy explaining that someone with MND had already contacted Dignitas but was 'watching what you're doing'; evidently I 'offered hope'. That message became my greatest stimulus to keep writing this book; there are many reasons why someone may want their life to end, but surely losing hope must be the saddest.

I find it not only incredibly sad but also incredibly frustrating that anyone would feel the pressure to die basically because they see no hope. No reasonable alternative. No other sensible option. I want them to see hope, alternatives, options. I want them to feel they have a *choice*. I passionately believe that what they do then is totally up to them, and I utterly support the right of people to choose death. But I equally strongly support their right to choose life.

Choice requires people to believe that there are serious alternatives for them to weigh up. Otherwise it's nothing more than a fait accompli – a sham ritual of decision-making whose outcome was never really in question. I want people to feel that they genuinely have options. And many, many people today simply feel that, in practice, there is effectively only one route open to them.

Consistently, I am told that people elect *not* to have feeding tubes or ventilation because, allegedly in full possession of the facts, they simply don't want to go through – or carry on going through – what they consider to be a torturous endgame. But I worry that many of the people making this decision may in fact be clinically depressed. Depression is a most awful illness, far worse than MND, and being depressed is not a good state in which to make balanced judgements about life or death. Yet that is not the main reason I worry.

To me, the ultimate tragedy about all this is *why* these people get depressed in the first place. 'Because they've been diagnosed with MND!' is an obvious reason. But that is *not* the reason as far as I can see. Instead, it's because the horrific way that the ultimate prognosis of MND is routinely portrayed can make death seem like a blessed reprieve from the alternative of suffering a living death – fully aware yet unable to move. Maybe for years.

And on top of that, some worry about the enormous emotional and care burden they'll place on their loved ones. They also worry about the *financial* burden. For some people, already scared of what they're facing, it almost feels like an act of love to sacrifice themselves for the good of those who remain.

In the past, that terrifying portrayal was largely true. But as a result, far too many have died of MND far too young. And far too often they've first been humiliated, terrified and destroyed, indignity by indignity, until even some of the strongest gave up hope, while others chose not to carry on because they didn't want their loved ones to suffer any more. Or they were scared that they couldn't afford the financial burden of choosing life. Or they were bullied by their medical team to 'let nature take its course'. Or they were simply never informed of the full set of options.

Who knows, back in the bad old days, what I myself might have chosen. But sure as heck, it's not the bad old days now. We are well into the twenty-first century. Cutting-edge hi-tech continues to exponentially rocket upwards in power. Those of us with MND can, if we choose, have a fulfilling and exciting future ahead of us. It is already possible to be semi-independent. And productive. And to have fun. And the hi-tech enabling these things is becoming even more amazing. Why would anyone want to miss out on all that?

Of course, to change the world, technology alone is not enough. If we only come up with brilliant ideas, we fail. If we only build amazing proofs of concept, we fail. If we make hi-tech available that people don't access (they don't want it, aren't offered it, can't afford it, don't live long enough), we fail.

But if we forever change the world so that anyone, if they choose to, can THRIVE even with extreme disability – only then do we succeed.

So, in addition to devising jaw-dropping hi-tech, we also need to change attitudes. We need to view research as part of a major change intervention, stimulate awareness via conventional media and social media, and as necessary lobby governments and healthcare communities.

To *keep* changing the world, those of us researching in

this field must constantly regenerate and evolve what we do and how we do it, constantly pushing back the frontiers of what is possible, constantly positioning ourselves at the cutting edge of applying hi-tech to MND and extreme disability, constantly leveraging Moore's Law to translate research into user-tools, constantly striving to deliver more and more support to *everyone* who dreams of breaking free from the straitjacket they find themselves in. I know it sounds hackneyed, but if not now – *when*? If not us – *who*?

Hello, Peter,

It's Julian's mum. We want to thank you for your intervention. Attitudes towards surgical procedures for Julian have changed dramatically.

Almost every clinician visiting Julian in hospital has mentioned your name with great excitement.

There is to be a review and forward plan for potential laryngectomy or tracheostomy for Julian and his surgeons are going to liaise with Plymouth with regard to the procedure, and there is even the possibility of Julian having surgery there if they feel they do not have sufficient expertise to do it [here].

This is all great news for us and I am so pleased my daughter contacted you.

I also believe you have broken down barriers to take your Right to THRIVE campaign forward.

I sense a wish by the practitioners [here] to get involved in the innovative and pioneering interventions for people with ALS/ MND and the attitude of no hope has completely reversed, which I see as a great step forward.

I do not have words to express how grateful I am for your help and support without which Julian would have had very little time left on this Earth. He wants to thrive *and you have facilitated that* right.

Someone said to me a few days ago that people were not allowed the right to die but in this situation they could be denied the right to live. I thought that was a very poignant comment.

My heart aches for people who do not have the knowledge to fight the system. I am so very glad we made contact with you. You are our inspiration and you have given us hope.

Once again, my heartfelt thanks. We can never thank you enough for your intervention. Please let me know if there is anything at all I can do to promote your wonderfully inspiring work.

Guardians of the Flame

With my biological voice now incredibly weak, I asked the almost thirty people currently sitting in the largest room of the DXC Innovation Centre in London to gather around me as close as possible. Of all the megacorporations I'd spoken with, only two had displayed the courage and leadership (or was it the commitment and social responsibility?) to actually turn up. But two megacorporations was more than enough for this two-day work session to kick off the Foundation.

DXC was there, of course. But also represented was the amazing Intel – the same team that worked with Stephen Hawking. And then there was CereProc, and Esther's colleagues, and the numerous experts involved in creating the avatar, and a smattering of other individuals key to the Foundation.

We were at the start of the first day. The schedule was full and we'd be working long hours – 8 a.m. till 8 p.m. both days – and I needed to get things going. But, even more importantly, I needed to set the Foundation on the best possible course for the foreseeable future.

Everyone in the room dutifully wheeled their chairs in my direction. I encouraged them to cluster together even tighter, until eventually any one of them could have heard me whisper. And then I began:

'This is the very first gathering of The Scott-Morgan Foundation! And these next two days are also the very last time many of you will hear my biological voice. So, I thought I'd mark these milestones by opening a portal that links both the past and the future.

'In August 1984, thirty-five years ago almost to the day, my very first book was published: *The Robotics Revolution*. I'd had a huge battle with my editor (and the professor reviewing it) over the ending. In it, I predicted the far future. But I was dumb enough to predict a future that was close enough that I'd still be alive to be proved wrong! I chose half a century. At the time, that felt a *very* long way away.'

Just as I had described to Francis, I went on to explain my fear that humanity had come to the fork in the road that I'd long ago predicted, and was now unwittingly choosing the wrong route, heading towards stand-alone AI (with the attendant inevitability that eventually humans would be left far behind) rather than human-AI collaboration that would let us do things that neither we nor AI could do alone.

'I never for a moment imagined that several decades later I would want to read out my musings. But, rereading the final paragraph, I do rather appreciate the cosmic irony of my cavalier use of the term 'all-too-vulnerable bodies' in the light of my inability now to even *hold* the book being quoted. So I'll ask Andrew to do the honours.'

On cue, my dutiful nephew had his moment in the spotlight, and read:

'If the path of "enhanced human" is followed, then it will be possible for mankind and robots to remain on the same evolutionary branch, rather than humanity watch the robots split away. In this way, mankind will one day be able to replace its all-too-vulnerable bodies with more permanent mechanisms, and use the supercomputers as intelligence amplifiers.'

I continued:

'Thirty-five years later, those of us in this room now get

the chance to nudge the future in exactly that direction, the ultimate example of what Jerry would call "being out in front where change is born" . . . You could *not* make it up! We are the key players in an origin story.

'But we all know, if this were a blockbuster movie – and one day it will be – this is the point where the going gets tough and the stakes get high. After all the wild talk, and unbridled excitement, and grand ideas, and shallow commitments, the true heroes of the origin story – the ones with the staying power and the focused talent and the burning passion to truly make a difference – they finally emerge. And they are *you*, along with a few other rebels who will be joining us in a few hours either physically or electronically.

'We are it! We are the true core of the Rebel Alliance.

'But this is a crucial time. Our rebellion against current reality, our mission to change what it means to be human, our vision of a world where collaborative AI liberates *all* of us to *thrive* – even people like me – all of that comes to an end if we don't succeed in our next steps. For a very simple reason.

'There's no one else out there in a position to do what we can. No one else out there is even *trying* to do what we are attempting. On the contrary, the status quo that we are pushing against is very strong. The default future that we are striving to change has already largely been written with the *wrong* ending. The *real* reason that the going will now get tougher and the stakes will get higher is because our biggest challenges are not technological – immense though those are – they are *psychological*.

'We've all heard the rumblings of a backlash against AI stirred up by the news media and Hollywood and concerned celebrities and members of the public. We need to create an alternative. A way out. Dare I say it, a New Hope.

'Yet even as we've taken our first steps in that direction, we've already seen some parts of the Establishment get cold feet, uncomfortable about technology they don't understand. We've already seen some corporations and individuals go silent, stand back, choose to not get publicly involved, to wait to see what happens, uncomfortable about not fully being in control, concerned about reputational risk, confused about what's in it for them.

'And I understand them. I get it. They are what they are.

'But *we* are what *we* are. By a process of natural selection, every rebel in this room has what it takes to change the world. All of you are squeezing precious time out of ongoing commercial projects to donate your amazing talents to work with others for our common cause. And some – not least the wonderful Jerry and DXC – have at the *corporate* level had the guts and leadership to publicly stand beside us in our struggle to break society free from the status quo and to prove that, *whatever* the universe throws at us, we can all *thrive* on change by embracing it, leveraging it and leading it.

'Starting today and tomorrow, we get the chance to demonstrate a different way forward, a non-threatening way, a safer way – as Jerry wrote recently: "AI deployed alongside people who can apply context, common sense and creativity." That's what we're really doing here. We're proving a point. We're not talking about it. We're *doing* it! Showing it. Making it work.

'And just as impressively, we are today encoding that visionary can-do attitude in a way that should ensure that it lasts far longer than any of us. We are embedding it into the design of a unique research body structured as a charitable foundation.

'The official goal of our philanthropic foundation is just a rephrasing of the dream that brought us all together in the first place, namely: research into the ethical use of artificial intelligence, virtual reality, augmented reality, robotics and other high-technology systems to enhance the capabilities and wellbeing of those restricted by age, ill health, disability or other physical or mental disadvantage.

'Within that context, the Foundation is being formed to very explicitly and very publicly harness the power of innovation to *thrive* on change – *even* when the change is induced by MND.

'I have exactly eight weeks left before I will never talk naturally again, and I'll be dependent on whatever the Foundation's research comes up with. But what we are doing over the next two days and beyond is not just hugely important to me, it's crucial for everyone with extreme disability and, ultimately, it's crucial for all of us.

'Of course, the Foundation will be a beacon of hope to those who currently have none. Yet it will also enlighten the progress and the direction of AI worldwide. And the brighter our flame shines, the more others will want to join us to help it to shine even brighter. But never forget, *we* are the guardians of the flame. And forever at the centre of that flame is humanness, humanity.

'I know that over the next two days there will be times of great frustration. I know that we'll horrify ourselves by discovering vast gaps in our thinking. And I know that we'll all get tired and make mistakes. But that's OK. This is the beginning of a wonderful journey. We don't have to get everything right. We just need to take our first bold steps towards a bright future. I am immensely proud to be on this journey with you. And, from the bottom of my heart, thank you for choosing to travel alongside me.'

'She's quite amazing!' Back in our hotel room, Francis was pronouncing judgement about Lama, an Intel fellow and director of one of their AI labs. 'She held court for two solid hours and everyone was hanging on her every word!'

'Remember, her team developed everything for Stephen Hawking. She worked with him for seven years.'

We'd first met Lama a few months earlier; she'd flown in from California and we'd spent the day together in London. Near the end of the day, the Channel 4 film crew asked her for an interview. Halfway through, she'd accidently started referring to me as 'Stephen' – which I'd considered the greatest possible compliment she could bestow.

'Thank you so much for my catheter, by the way!'

This had just popped into my mind and I suddenly worried that, in the rush, my thanks earlier in the day had been insufficiently effusive. I'd been in one of my back-to-back meetings when I'd realized that my bladder was full. This was supposed to be impossible. After all, I'd been replumbed a year earlier specifically to preclude any such incident. It was a matter of seconds to conclude that my catheter (output no.1) was blocked.

This was unfortunate. However, very soon, it also became very painful as my bladder overfilled and, like every over-inflated balloon, risked bursting. This would be even more unfortunate, I reasoned, and I reluctantly brought my meeting to a premature conclusion and advised Francis of my predicament.

A blocked suprapubic catheter is cause for an emergency trip to hospital, where in principle a doctor or nurse can replace it, though only a few have ever had the requisite training. Fortunately, we had anticipated this eventuality and had, highly unusually, got Francis trained and signed off to replace one himself. Crushed into an overheated disabled toilet, he took full advantage of CHARLIE's ability to become completely horizontal and used my legs as a table, and proceeded to execute a perfect recatheterization using selected contents from the emergency bags we always travelled with. Ten minutes later, I'd been in my next meeting.

'That's what I do,' he self-deprecatingly but accurately replied. Then he made a conversational course correction: 'That Steve's great, isn't he?' This was our new friend from DXC London who'd become such a supporter that he'd tracked down an ancient copy of *The Robotics Revolution* for himself. 'And Jerry was talking about exploring some form of strategic partnership, wasn't he?'

'It would be absolutely amazing! If we can get proper sponsorship for the Foundation, our vision will be safe. We can relax. If someone commits at the corporate level, then *everything* changes and none of the hard work and sacrifices you and I've made will have been wasted. They would *join* us as guardians of the flame –'

'Oh, bugger!'

'What?'

'I forgot to point out to the board just how diverse we are!'

'How do you mean?'

'Well, there are eight of us, right? And normally that would mean eight straight white men. Whereas we've only got one! How cool is that? I was going to suggest that we refer to him in the minutes as "Token" . . .'

Any Last Words?

'It's a huge luxury to be able to choose the very last words I'll ever utter,' I enthused to Anthony. 'Everyone should be so lucky!'

With my laryngectomy now imminent, we'd broken the habit of half a century and – rather than wait to meet in person or resort to written communication – had updated our friendship to the twenty-first century and were now using WhatsApp for a video call. I was in my study. It being early Sunday morning in Chicago, Anthony was in his luxurious forty-eighth-floor condo overlooking (as a vertigo-inducing flick of his mobile demonstrated) a breathtaking cityscape with Lake Michigan beyond. He'd ended up living a few blocks away from where the two of us had stayed in 1976.

'Have you decided what last words they'll be?'

'Absolutely! They're pretty obvious . . .'

Either they were too obvious to waste time on or Anthony was too respectful to enquire further.

'Your computer voice singing "Pure Imagination" was amazing. It was just like I remember your singing voice – good but untrained.'

'Francis says I sound like Angela Lansbury singing "Beauty and the Beast" . . .'

'Rex Harrison in *My Fair Lady*, surely! Anyway, tell me how you're doing.'

I'd totally forgotten that last time Anthony had seen me I'd been symptom free; now I could hardly speak and he could see that I was immobile in a wheelchair. Superlatively brought up in Wimbledon as he had been, he'd naturally made no direct mention of my deteriorated condition.

'Well, I have to say, I'm discovering that there are *huge* upsides to being paralysed. Especially when I'm being massaged – I have almost a couple of hours a day. It's brilliant! And as for being showered, even a pharaoh wouldn't be as coddled. It's like checking in to the most luxurious spa planet in the Milky Way.'

We chatted about how Francis had designed a completely new decked garden that allowed me to enjoy again the feeling of sitting in the shade of a palm tree. Then there was the lady who wanted to stage an opera using performers with artificial voices. And the fact that I'd almost finished my book.

'It gets better: we've just sold the film rights to the production company that won the Best Picture Oscar for *The King's Speech*.'

'Wow! But do they *realize* that the world's first ever proper cyborg is actually gay?'

We carried on chatting, just as inanely and importantly as we had on that sunny May afternoon on another planet and a different timeline when I was sixteen.

'We'll talk again when I'm a cyborg!'

'Give 'em hell! Lots of love . . .'

It was a long night. I'd taken summer for granted, and now the October dusk seemed impossibly early, and I was awake hours before a rather bleak dawn finally made its unenthusiastic arrival. Then again, the unfamiliar bed and noises meant I was awake anyway. Thinking. Listening to the hushed night-time regime of the hospital ward, punctuated by regular visits to my side room by a nurse checking my vital signs. When occasionally I drifted off to sleep, I woke soon after with an urgent jolt of adrenaline, remembering what was about to happen. I preferred to stay awake; I'd have plenty of time to sleep later.

It was the first time I'd had a chance to think – I mean *really* think – for a couple of years. Not since my self-diagnosis. Now I had absolutely nothing to do but wait. And think. Before everything got busy again preparing for my forthcoming meeting with my ENT surgeon, Philip. I felt surprisingly calm about it, given that he was set to cut my throat from ear to ear and excise my voice box.

I found myself reflecting on just what Peter's Philosophy of Life was, what belief system my subconscious took refuge in at times of anxiety, what (in the absence of religion) I *actually* had faith in.

This was not a new train of thought; I'd decoded the relevant Unwritten Rules decades ago. I just hadn't thought about them explicitly for ages. Now, with no work to do, I lay back in my hospital bed and contemplated the cosmos.

At its absolute core, navigating existence is surprisingly simple – you merely need to align yourself with the very few *all-powerful* Unwritten Rules, the ones that dominate all the others, the ones that run our universe. And fortunately, there are really only *three* such important Rules of the Universe; everything else is detail:

1. Science is the only route to Magic.
2. Humans matter because they Break the Rules.
3. Love – eventually – Conquers All.

Thanks to the Law of Logic and Love (the one I'd explained to Anthony on a Greyhound that dictates 'true love wins over logic every time'), the Third Rule of the Universe is the most important, the most powerful, the Rule that rules them all.

I drew immense strength and comfort from these three rules. Indeed, revisiting them after years of neglect, I realized that I had in the intervening time developed an *exceptionally* deep faith in them.

Firstly, I reminded myself, I had total faith in science. Not that it was always right, but that it would keep improving. And, unlike dogma, would steadily become less and less wrong. And, I told myself, if anything could improve my life with MND, *it* could. Far more wonderfully, as science pushed into new frontiers, the phenomena it explored became increasingly fantastical, improbable, magical. But they were real – or at least as real as anything else. Just as I'd explained to my art master, that's what I'd always adored about science. Just because you discover how magic works, that doesn't stop it being magic.

Secondly, I had huge faith in humanity. Not that everyone was kind – I knew they weren't. But that despite the vile cruelty and repellent inhumanity of some, our species as a whole was quite magical and extraordinary and indomitable because we alone in this corner of the galaxy deliberately break the rules. That makes us special. Important. We matter. Far more worryingly, we *might* be alone. Other sentient life – if there *is* any elsewhere in the cosmos – might not break the rules as habitually, as creatively, as impactfully as we do as a species. The possibility that we are crucially unique carries with it an enormous obligation to make a difference – to the universe. And despite our supreme stupidity sometimes, together we'll work things out. It's what we do.

Thirdly – in flagrant contravention, I knew, of the trite, overused Hollywood trope of the 'cold, passionless scientist' – I had an unshakeable faith in the power of love. Not that being touchy-feely to everyone would solve anything – I knew it wouldn't. But that when all else failed, when there was no more hope, when against all the odds no sane animal or rigorously logical robot would ever carry on, then – despite everything – the irrational, stubborn, absurd, self-sacrificing, blind, unstoppable, magnificent, all-conquering, unconditional love of one frail human being could reveal itself to be one of the most formidable forces in the universe.

Just because I knew that this Third Rule was not a miracle but ultimately came down to hormones and genetics and neural networks and complexity theory did *not* stop it being achingly wonderful and glorious. That's where it linked back to the Second and First Rules: sometimes, only love was mad enough and brave enough to break the rules to their very foundations; sometimes, only love – real as it was – could conjure pure magic.

'This might make you feel a bit floaty.'

Maree was ready to anaesthetize me. But, as agreed, she'd paused the countdown. This was the final go/no-go decision. It was also her prearranged euphemism for 'any last words?' The clock would only resume upon my signal. Highly unusually, the antechamber to the operating room where I was about to be anaesthetized was crammed full of people. This was partly because it was in essence a glorified corridor and partly because, despite its compactness, it was now accommodating Maree, her assistant, me on a hospital gurney, Francis and a Channel 4 film crew – with me in a gown, everyone apart from Francis dressed in scrubs, and all of us surrounded by impressive-looking surgical equipment.

Francis took his cue from Maree and moved to my side. We hadn't discussed it, but I'd sort of expected him to say nothing, just wait for me to say something. Instead, he began:

'People are thinking the last words are gonna be –'

'Four words . . .'

I hadn't meant to interrupt him, but I was suddenly worried that the pre-meds might slur my speech if I left things too late. And these were the four most important words I would ever speak.

Francis leaned in close. This was the only part of the day that I'd thought much about, interminably, for more than a year. Rehearsed, really. For some rather illogical and utterly human reason, the repetition had only increased the significance rather than diminished it, as I'd hoped. As a result, this moment had grown and grown and grown in importance, until by now it dominated how my brain viewed its whole existence and had been elevated into the ultimate watershed: pre-laryngectomy and post-laryngectomy; before and after losing any means of communicating conventionally; Peter and Peter 2.0.

I wasn't in any way scared of the operation; I had no doubts whatever that it was the right course to take and that this was the ideal time to take it. My biological voice was getting difficult to understand, and my synthetic voice was already far more like me than I was, so substituting new for old was an unequivocal upgrade. And, far more important, my laryngectomy was my passport to a potentially long lifespan; almost exactly on the date that statistically I was scheduled to die, I was now scheduled to take steps to live indefinitely. I was pushing back. I was taking control of my life. It was a glorious new beginning.

I'm just not very good at goodbyes.

And despite every shred of logic and common sense that I could throw at the situation, I *knew* that this could not be dismissed as just one more in a series of myriad goodbyes that together made up the 'long goodbye' that was MND. I'd already registered my last bath, the last time I'd climbed the stairs, the last time I'd walked on Dartmoor, the last time I'd walked anywhere, the last time I'd been able to get myself out of bed, the last Christmas dinner that I could eat unaided, my last signature, the last time I could type even with one finger, the last time I could surf the internet by myself, the last time I could hug someone.

And then, in the run-up to my laryngectomy, an increasingly rapid-fire onslaught of the last time I'd sing happy birthday, the last time I'd smell the evocative salt of the sea, the last time I'd speak directly with one friend or family member after another, the last time I'd wish Francis goodnight. And now this. This *was* significant. It was *not* like all the others. It was the ending a major chapter of my life, just as much as it was the beginning of an exciting new one. In my innermost being, I *knew* that.

I also knew that all I now needed to do to restart the mission launch that would propel me into this new, unexplored, exciting parallel universe was to utter four simple words, the last words I would ever breathe. Then Francis would kiss me. Maree would put an oxygen mask over my face, tell me she was about to put me under, and seconds later I would sense the slightly woozy signal that my irreversible journey to another world had begun. All I had to do now was to trigger that final countdown. And yet I hesitated.

I knew that the next six months would be utterly horrible. I knew that I would loathe suddenly not being able to speak, on top of already no longer being able to move. I knew that with air no longer flowing through my nostrils I would lose my sense of smell, and food (which I should still be able to eat for a few more months, provided it was soft enough) would taste as boring as during a cold. I knew that I would feel claustrophobic, vulnerable, impotent. I knew that hardly any of the hi-tech I depended on to climb out of this valley of despair was yet available. I knew that even when it was, it would crash, have bugs, not work properly. I knew that there would be times when I would feel pitifully sorry for myself.

I also knew that the next two years would be increasingly exciting. I knew that my verbal spontaneity systems would begin to work properly, begin to accurately predict what I'd started to spell out, begin to listen to (even watch) what was going on and make good suggestions for how I should respond. I knew that my personality retention systems would also begin to work properly, my low-resolution avatar would move in real time, display the right sort of emotion, smile a lot, show people that I cared about them. I knew that I'd be able to give complex speeches again, and I'd be able to sing again, and I'd be able to create art again. I knew that my book would be published, the documentary would be broadcast, maybe even a movie would be in production. I knew that the Foundation would go from strength to strength as word got out of what we were attempting. I knew that even after two years, this would still only be the beginning.

I knew that the next two decades would be utterly amazing. I knew that more and more people would benefit from the Foundation's research – not just those facing the challenges brought about by extreme disability or old age but also those with the curiosity and courage to simply want to be different. I knew that AI systems would double in power every two years, and that therefore *my* powers would double every two years, accelerating exponentially. I knew that virtual reality, fuelled by the computer games industry, would reach such levels of staggering realism that people like me – cyborgs like me, with very high-resolution avatars – would at last be free again, unstoppable again, young again, even immortal. I knew that in maybe ten to fifteen years brain-computer interfaces would finally surpass the speed and accuracy of eye-tracking systems, and from then on those of us who wanted to would be able to link ourselves directly with AI, work seamlessly with AI, *meld* with AI. I knew that, if I could just help nudge the future in the right direction, everything I had dreamed of as a teenager was on the brink of becoming true; I had only to survive long enough to be part of it, to help make it happen, to help rewrite the future.

It is the birthright of every cyborg to be able to change the universe. And I'd been preparing for this journey since I was sixteen years old.

So I triggered the final countdown.

Slowly.

As precisely as possible.

Fighting my inability to speak clearly:

'I – love – you . . .'

And then the only possible choice for the single last word ever to pass my lips:

'. . . Francis.'

Twenty-one Years Later

Salania

'Highcliff can help me work here just as well as in my bedroom.' He was mid-twenties, slim, and taller than his great-uncle who – given this was the conservatory – was rather eccentrically dressed in a tight-fitting VR suit. 'Anyway,' he put on the mock-reassuring persona he'd used as a teenager when justifying why he should climb on the cliff at the end of the garden, 'what could *possibly* go wrong?' He pulled his impish grin, and that settled it.

Francis smiled his thanks, then took a protective look at what seemed like a wizened waxwork sitting with its eyes and ears obscured by a wraparound visor.

'He's fully immersed, so he won't be able to hear you.' As explanation: 'He said he had a lot to show me.'

'It's wonderful to have him home.'

'It *is!* And I must hurry to him . . .' Then, in a slightly louder voice, directed nowhere in particular: 'Highcliff?'

'Yes, Francis.' It was long-time friend Jerry "the AI Wizard", now also the voice of the AI that ran the house.

'Can you take me to the library please?'

'Your wish is my command!' – with a laugh in its voice. Part of the glass wall to the open-plan living area slid open and a sleek autonomous chair whisked in.

'Oh, he's getting good!' Ollie murmured as he helped Francis shuffle over and with a sigh of comfort sit down.

The chair smoothly accelerated toward a large glass lift in the distance. A minute later, Francis emerged into the highest, most futuristic room of the house. There was an enormous viewscreen showing a live-feed of an exquisite alien landscape, an entire wall of signed first editions of some of the last books ever printed, an eclectic collection of artifacts from across time and space. Francis saw only a magical-looking map on the wall. The chair glided in front of it. He picked up a visor from an old sea chest. The window-wall overlooking Tor Bay slowly went opaque.

'Highcliff: Salania, full immersion please!'

'Welcome back!' The voice was deep and friendly but sounded as if it could very easily turn into a growl. 'Where would you like to go?'

As usual, the portal to Salania was a facsimile of the original map showing the three realms and surrounding lands, complete with annotations.

'Where is he?' Instantly, Francis felt the familiar rush as he heard the voice compensator make him sound young and energetic again.

'The Archmage is in the Citadel.'

'Then Lusion it is!'

At once, the map became three-dimensional and tilted away until Francis was flying high above a real landscape, accelerating at impossible speed, a blur, and then he was riding his white stallion, Mist, with the gleaming white walls of the Citadel of Lusion right in front of him, the vast plains of Fyson in his peripheral vision. He knew that he was in his early twenties again.

There was something about the light in Salania that Francis always found uplifting. It was slightly yellower than on Earth. And the birdsong was clearer, and the colours made everything look like a Maxfield Parrish painting – just as Peter had always wanted. He looked down to his right, noted the fine white cotton of his balloon-sleeved shirt, the dark royal blue of his leather kilt, and felt the thrill of seeing his tanned and muscled thighs astride the broad saddle. It was extraordinary how shallow the human brain was. As he looked appreciatively at his restored legs, he felt his self-esteem returning. He utterly loathed the thin and wrinkled, poorly functioning legs he had on Earth. Here, now, he was strong and vibrant again.

He caught a flash of movement. There was loyal Tangbone padding along beside. He was like an Irish wolfhound – but larger, with brilliant light-blue eyes. And he'd never die. The perfect canine companion.

'Hi, Tang!' They were about to cross on to the long narrow stone bridge across the dry moat. The portal always introduced a destination gradually to allow time to adjust. 'We're in a hurry.'

Mist broke into a fast trot, with Tangbone lolloping along with an easy stride. High on the battlements, one of the Watch signalled the arrival of a prince with five double blasts on his horn. Only the King would have got six. Years ago, Francis had stopped telling himself that it was all silly, all an illusion. The truth was that his subconscious had begged him to willingly suspend disbelief. Just go with it. Enjoy it. And these days, that was exactly what he did. He assumed, but wasn't sure, that the guards were all algorithms. But being welcomed like this nevertheless always set his heart racing.

Seconds later, Mist clattered across the huge ironwood drawbridge; there was another clatter as the four guards stationed at the gate slammed their metal-tipped pikes to the flagstones in salute, and a different clatter as Mist moved on to cobbles under the portcullis, through the gatehouse archway and into the Gaurderian Parade, a space that Peter insisted was large enough for a thousand mounted Protectors.

It was soon after dawn. The twenty-nine-hour Salanian day meant it always felt a toss-up what time it was when a user arrived or left. That way, Peter had said, it didn't matter what time zone people entered the portal from. As it was, the Parade was almost empty, apart from what looked like two men (one of them bare-chested) on the back of the same jet-black horse galloping across from the far corner. Francis realized they were charging directly towards him. This was new. As they got closer, it became clear that it was one man on the back of a centaur. That really *was* new: the online centaur community didn't like the symbolism of someone shown on their back – it was a centaur rights issue.

Then he could see that it was Aril who was riding bareback. Dressed like a man, as usual, she was mounted on a young centaur – if he'd been human, he would have only been in his late teens. To be fair, Francis thought, that was Aril's apparent age too – though back on Earth she must be as old as Methuselah by now. They skidded nearly to a halt, veered around and fast-trotted on Mist's left side, in the direction of the entrance to the Grand Hall.

'Fuck protocol! How is he? Tang told me he's just got back from hospital.' She of all people knew that everyone was supposed to stay in character, not break the illusion, not remind people that there was another world in which they were not as fortunate. She had once told Francis that she considered it to have been one of the Foundation's greatest achievements to trigger the creation of an ever-richer multiverse of virtual worlds in which *anyone* could thrive – whatever their disability and however narrow their means of communicating. She'd been one of the first users. And from the start she'd been very vocal in insisting that it was bad etiquette to mention what she called 'Sadworld' in any Salanian conversation. Right now, she didn't seem to care.

'Aren't you in Seoul at the moment?' Francis didn't care either.

'Yes. I haven't looked, but it must be the middle of the night. But I needed to see you.' She suddenly smiled and looked girlish, despite her mannish attire. 'And to see Iridon, of course!'

Aril reached forward and placed a proprietorial hand on the centaur's neck, caressing his mane of raven hair. As she did so, he looked back, smiling, revealing beautiful green eyes in an equally beautiful face. Then, never changing his torso's position relative to Mist's neck, and somehow never slowing his progress toward the Great Hall, he completely turned around until he was fast-trotting backwards. Even Aril looked mildly surprised, and she was *always* so cool.

Iridon bowed his head, clasped his fist to his bare chest and looked up again.

'It is an honour to finally meet you, my liege.'

Like all centaurs, his voice sounded impossibly low.

'And you, Iridon. Welcome to the Protectors. When did you join?'

'At the triple moon, my liege.'

'He lets me ride him . . .' Aril explained rather redundantly.

Padding along beside Mist, Tangbone spoke up for the first time, his voice still a deep, friendly growl despite the exercise. 'Well, there's a surprise!' He'd got canine sarcasm down to a T. He looked up at Francis with his doggy eyebrows raised. 'She's been riding anything with a pulse since she was twelve.'

'I heard that, you embittered old fleabag!'

With perfect manners, Iridon turned his stallion body around again until he was once more trotting in the direction he was facing, with his muscled back to the conversation that he could now at least pretend not to overhear.

\<Is he AI?\>

Francis used privacy mode to ask Aril the question telepathically to avoid any risk of Iridon being embarrassed.

\<No, I think he's real.\>

\<I resent your insinuation that *I'm* not real!\> growled Tangbone, who as the prince's familiar had been coded long before most of Salania had even existed.

\<Sorry, Tang, you know what I mean. No offence . . . \>

\<None taken. By the way, have you told him you're a man?\>

\<Ouch! You must be riddled with bugs today to make you so nasty! And no, not that it's any of your business, I haven't told him. And I haven't asked him if he's a woman. Or what age he is. Or whether he's AI. I really don't want to know. At least for now. I'm just enjoying getting to know Iridon for whomever he wants to be.\>

She paused, then, with a look of utter innocence, added:

\<And the sex is *great*!\>

\<*Far* too much information!\> Tangbone growled. \<Anyway, even in Salania, I suspect that isn't physically possible!\>

\<Don't you believe it! He's built like a –\>

\<*Shut up!*\>

They'd arrived at the Great Hall. With eight hooves now silent, faint sounds of the Citadel waking up drifted into the Parade. Aril turned in the saddle and looked straight at Avalon.

'As an old friend, I'll ask again: how is he?'

Even though he had only been there for a few minutes, Salania had already begun to work its magic. The facts remained the same, but for some reason those same facts now felt as if they referred to another time and place comfortably distant. It felt a bit easier to discuss the truth here.

'They can't do any more for him in hospital – so he's come home.'

'Oh! I'm so sorry! Have they . . .' she paused, but never broke her direct gaze, '. . . given any idea of timing?'

'Not really. Days? Months? But his brain is unaffected. I'm just going to see him now.'

'Then I won't keep you a second longer. Run! Give him my love!'

'Of course I will.'

Avalon swung down from his saddle and, with Tangbone by his side, strode to the huge doors that, even as he approached, two guards in their crimson-leather kilts pushed their shoulders against to open. It felt so good to stride rather than shuffle. He looked back. Iridon briefly touched Mist's forehead with his hand, then trotted away with Aril looking grim and Mist obediently following.

The Great Hall was utterly enormous, designed to amaze and inspire. On another timeline, it had been the backdrop for Rahylan's near-fatal triumph at the Warlock Games. It was such a pivotal episode in *The Ballad of Avalon* – the back-story behind the original map – that Peter had persuaded Francis to join him in creating, in a mind-warpingly intense computer session, the final round of the competition for real, as Avalon and Rahylan, with all the jeopardy of the original.

Then, with that timeline safely stored, they'd both shapeshifted into different bodies and gone back in time to relive the whole thing, participating as members of the audience. Even more impressive (from Francis's point of view) was that in his disguise he'd been able to create a crucial diversion and *revise* the version of Salanian history that then got archived.

All that now felt irrelevant. This time, Avalon ran through

the enormous room, focusing only on the Great Staircase ahead. It was so wonderful to run again. To be twenty-two again. He reached the stairs and started bounding up, two at a time, racing Tangbone, intuitively pushing down on the pommel of his broadsword, Febrion, to raise the scabbard so that the tip didn't catch on the steps. He couldn't even remember how long it was since the adapted gaming software had become that sophisticated.

With a burst of speed, Tangbone made it to the landing first. But only just. And there it was: one set of double doors; two possible rooms beyond. The first was merely the Great Audience Chamber, which anyone could enter. The second was the Archmage's Library, Peter's Library, which wasn't even there unless you were supposed to go in. Francis hesitated, as he always did, with his hands on the two large phoenix doorknobs, wondering if this might be the first time the Library wasn't there. Then he pushed the doors open.

The Library wasn't really a room at all. It was more a space. It was a perfect cube, with each side more than ten long strides across. The face of the cube opposite the door was an enormous annotated map of Salania, identical to Peter's original except that the dolphins in the Bay of Maridorn were playing in the sea, the flag was fluttering on the Tower of Tiros, and Francis could just see the four guards he'd ridden past when he entered the Citadel.

Each of the remaining faces of the cubic Library was a floor, even the ceiling, on which a very large black puma was strolling about upside-down. The cat flicked his light-blue eyes to the door, then bounded along the 'ceiling' and at the edge leaped across to the 'wall', continued to bound head first down to the 'floor' on which the visitors were entering his domain, leaped down to that and, suddenly adopting a more dignified gait, padded over to them.

'Hi, Charlie!' Francis said to his old friend.

'Welcome back! Both of yuh.' The oversized cat's soft Scottish burr was as reassuring and fantastical as ever. He nodded an acknowledgment at Tangbone.

If Rahylan's familiar was here, Francis reasoned, then Peter had to be very close. But the Library looked empty. Avalon looked left, right, up, as Tangbone and Charlie briefly touched their foreheads together. Looking around, Francis found the Library as unreasonable to pigeonhole as ever. It was timeless. More accurately, it was *all* times. And many places. Some surfaces were old flagstones or well-worn ancient floorboards. Others were impossibly perfect crystal or highly polished metal. An outcrop of living rock covered in moss and ancient carved runes seemed to grow from one corner.

Some 'floors' acted as 'walls' holding bookcases of ancient tomes accessible if you stood in front of them on the correct 'floor'. There also, at eye height, was a large painting, *Metamorphosis*, Peter's first ever piece of cyborg art, but on this version the image impossibly zoomed in and out and flew around the emerging structures of the original. There was the collection of magazine covers about the Foundation. There was the movie poster. There was the fifties jukebox next to the alien-looking sculpture that didn't appear to be in the same space–time as its surroundings. Perched on it at a rakish angle was the denim hat signed by Nureyev. There was one of the framed T-shirts with a picture of Peter's early avatar and his first ever words after losing his voice: 'Peter 2.0 is now online.'

'Only Peter could have possibly come up with this,' Francis found himself thinking. Followed by a pre-shock of the grief yet to endure.

'I'm over here.'

The Phoenix Pyre

I watched Avalon wheel around. He looked as gorgeous as ever. I hadn't been able to see him in hi-res for weeks – unless I went back in time and watched him. But that wasn't the same; we couldn't interact. And in hospital, Francis had seemed very tired.

'There you are!'

I was standing on the 'wall' that had the entrance doors in it. As usual, I was barefoot, wearing only the brilliant-white ankle-length four-slit kilt of a warlock, held at the waist by the thin Avolean gold belt of the Archmage. My bare torso was just as it had been at my peak; my only jewellery was my wedding ring and my ankh.

As the doors closed of their own accord, I swung my hands to the banisters of a short flight of stairs at forty-five degrees between my 'floor' and Avalon's 'floor', slid down them and – without fully understanding quite how my avatar managed to execute such a feat of gymnastics – ended up standing in front of my husband.

'I really wasn't hiding! Charlie told me when you got here, so I was just finishing tying up a few loose ends. You must have *rushed* through the welcome scenes.'

'Of course I did! And I had to talk to Aril. She sends her love, by the way.'

I gave him a kiss, just as I had a few minutes before when he'd carefully positioned my immersive-VR visor and then hobbled to the study to put on a visor of his own. In truth, this kiss was better. For a start, it was on the lips, as it was meant to be. One of the hidden cruelties of MND was that, to minimize the risk of coughs and sneezes, for more than twenty years no one had been allowed to kiss me on the lips. Not even Francis. And now, since my latest brain-computer interface, I even got a basic sense of contact.

Also, for both him and me there was the reality of the face being kissed, the reminder of the true essence of who we really were, rather than the irrelevant yet dominant outer husk. I pulled back a little so I could look at him up close. Back on Earth, his eyes were still similar, more bloodshot, slightly less blue but still recognizable. Hardly anything else was. As for me, I rarely asked to see myself in an Earth mirror, and it was rarely offered, but I had by now lost so much muscle mass that when I did catch a glimpse, my brain rebelled against the imposter looking back at me, convinced that it was impossible for something that looked so long-dead to feel so alive. As Rahylan, however, I looked exactly as I had when Francis and I first met.

'Where do you want to go?' I asked out of the blue. 'I can't tell you how good it is to be back on a hyper-fast link!'

'You can tell me . . .'

We both laughed with the bond that came from using the same silly joke for sixty years.

'I've been stuck in this room for *sixteen days* while I was in hospital, with only Charlie for company after you went home each day. So, pick a place and time!'

'I don't care! Just so long as it's anywhere but Earth in 2040.'

'In that case, "a long time ago in a galaxy far away" it is.'

I strode (my avatar's default was *always* to stride) across to the hexagonal control console from Doctor Who's TARDIS circa 1970, flicked a few switches, pulled a lever and, to the tortured sounds of a time machine taking off with its brakes on, one of the 'walls' dissolved to reveal an epic view of space, glowing columns of gas and stars towering far overhead.

'Come outside . . .'

With the wall gone, we were able to walk out on to a huge semicircular terrace that stretched the whole width of the Library. It was one of my favourite viewing platforms – it was *so* incongruous for outer space. I'd modelled it on a cross between the balcony of a Venetian palazzo overlooking the Grand Canal and the balcony of a favourite villa on Lake Como. It was edged with a marble balustrade that incorporated planters, filled with fluorescent pseudo-flora coded by a tetraplegic artist in Moscow. As we slowly walked to one end of the long balustrade, where it emerged from the edge of the Library, Francis was looking up.

'It's beautiful.'

'I promised you "long ago in a galaxy far away". This is the Eagle Nebula at just the time civilization was beginning on Earth. The light from these stars set off on their journey thousands of years before the Egyptian pyramids were even conceived; it's only just reaching the solar system seven thousand years later.'

'Hmmm, that does rather put things in proportion.'

'They're called the Pillars of Creation.' I ran slightly ahead and leaped on to the top of the balustrade. I loved this bit of the software. 'Have a look below us.'

Francis peered over, instinctively pulled back, then tentatively looked over the edge again.

'Bloody hell!'

We were only a mile or so above the surface of an Earth-like planet, rushing over it at great speed. As I looked at my bare feet balanced on the thin balustrade, I felt a fleeting moment of vertigo even though I knew that my avatar algorithms wouldn't allow me to fall. I was totally safe.

'The point is, we don't know if there's sentient life out there or not. But there's a ginormous number of beautiful planets out there that *our* sentient life – humans – *could* live on. That might be the most important contribution that Earth ever makes to the universe. And ensuring that our gloriously indomitable, rule-breaking species survives long enough to spread out into the cosmos may be the most important contribution those of us alive today can ever make.'

He took the mention of life, just as I'd hoped.

'Darling, you know they aren't giving you very long . . .'

'That's what we need to talk about.'

'How should we spend whatever time we have left?'

'That's *exactly* what we need to talk about! There's a bit of a problem. Well, an opportunity really. Well, a bit of both.'

'We can sort out any problem. What is it?'

'We've a chance to rewrite the future of the universe.'

'Is *that* all? I always said to leave with a bang. Go for it!'

'It may be an absolutely terrible idea.'

'That's never stopped you! What's your Big Bang Theory?'

'It's the answer to what I call The Bright-Eyes Riddle.'

'Bright eyes? As in *Bright Eyes*? Our song? How romantic!'

'It felt right. To us, it's our love song when we met. But it's all about the Black Rabbit.' His face snapped to the present. 'Every line is querying the nature of death. So, my Bright-Eyes Riddle is: *If you're a cyborg, what happens when you die?*'

'Sweetheart . . .' Confused yet compassionate. 'Even cyborgs – even *you* – go the same as everyone! When you die, you *die*!'

'It's just that I'm not sure how much of me is going to die.'

'What?'

Against an astronomic backdrop, I set my avatar to slowly

walk the whole length of the balustrade, skipping over the occasional planter. Avalon walked by my side.

'It's like this. You remember how, before I first transitioned, I told you that I thought I might have solved the Uploading Problem?'

'Darling, I don't even remember the Uploading Problem.'

'Yes, you do!' I rather irrationally responded, given that, clearly, he did not. 'The reason the transporter on the USS *Enterprise* couldn't work, and they kept killing Kirk?'

'I haven't the slightest idea what you're talking about!'

'OK, a reminder. For decades, people have clung on to the idea that one day it will be possible to scan their brain and reproduce it digitally and therefore, having "uploaded" their brain on to silicon, they'll be able to cheat death and carry on indefinitely. But it won't work. They'll still die. It's the computer version that will live.'

'Of course! It's a copy, a duplicate. I remember all that.'

'In addition, it's a massively complex technological challenge to scan a living brain at sufficient resolution to recreate it as software. One day, yes. But not for a long time. So, even though computer technology these days is just about powerful enough to act like a human brain, we don't have a way to configure it. At least, not without destroying the biological original.'

'Oh, I'm remembering now: you said that whole approach was doomed to failure because they were answering the wrong question.'

'Exactly! They've all been asking how we can upload on to a computer. I think we should be asking how we can *become* a computer!'

'And I said that was stupid. It's all coming back now. You called it melding.'

'Melding, yes. It's my melding hypothesis. But I'm not sure that it *is* quite as stupid as it sounds.'

'OK, explain in words of one syllable.'

'Indubitably! Let me draw your attention to Exhibit A.' I paused for a moment near the centre of the balustrade, twirled to face Avalon and, with a flourish, pointed my hands at myself. '*I* am Exhibit A!'

'Of course, you are . . .'

'Well, twenty-odd years ago I began melding with humancentric-AI systems. We forget, but originally it was simply in order to speak and express emotion before people fell asleep waiting for me to say anything.'

'That feels an age away.'

'It does. None of this' – I waved my arms at the Library, pivoted to include the cosmos, and continued my stroll along the balustrade – 'was vaguely possible back then. Why?' I knew that Francis knew the answer.

'Moore's Law!'

'Moore's Law. The power of my AI systems has doubled every two years. $2 - 4 - 8 - 16 - 32 -$'

'Yes, yes, I know!'

'Yes, but this is when it gets mindboggling: $64 - 128 - 256 - 512 - 1024$! That is what twenty years of Moore's Law means. I am now a *thousand* times more powerful than when I first transitioned.'

'You promised that would make you funnier. Or at least more intelligent.'

'I am, it's just that you're getting harder to impress. But my point is . . .'

'Yes, what is your point?'

'My point *is* that ever since my latest brain–computer inter-face, five years ago, my biological brain has been able to interact with all the self-learning AI systems, and – I've mentioned this several times – it's become more and more difficult for me to distinguish which is which.'

'I thought you said it was like having a friend inside your head who could communicate with you telepathically.'

'It was, to start with. But as my AI has got better and better at second-guessing what I want to say, how I want to move, what I want to do, it's become less and less clear to me where a particular idea originates.'

'But how's that even possible?'

'Consciousness, self-awareness, doesn't reside in any one part of the brain. It emerges as a sort of side effect of all the processing that's going on. That's why they finally brought in legislation to protect AI that passes the revised Turing test.'

'If it's as clever as a human, and it claims to be self-aware, then it could well be telling the truth, so you have to treat it accordingly,' Francis paraphrased.

'Precisely! So, my melding hypothesis is that because "AI me" and "biological me" have been integrating better and better, and AI me is getting better and better at second-guessing and mimicking biological me, and AI me is getting more and more powerful whereas biological me is getting older and more forgetful –'

'I hadn't noticed . . .'

'Well, I have a sneaking suspicion that AI me is increasingly taking up the strain. And more and more of my consciousness is being generated as a side effect of AI me rather than biological me.'

'That must feel really weird!'

'But that's the point – it *doesn't*! It just feels like thinking.'

'Bloody hell! And you're wondering what's going to happen when biological you dies?'

'Exactly!'

'Fucking hell, you're serious!'

'*Deadly* serious.' I grinned, mildly self-satisfied that I could still pun even as I was finally dying.

'You think you might survive?'

'No. But I think that *some* of me might survive. And it might be recognizable. In due course, it might even be funnier and more intelligent.'

'Well, *now* you're talking!'

I'd reached the far end of the balustrade. I hopped down and rested my hands on Avalon's shoulders.

'Darling, there's something very important for you to think through. When biological me dies, do you *really* want AI me around? Even if AI me survives at all, I might be weird, or appear mentally ill, or stupid. Or I might be OK, but then get cleverer and cleverer. I might end up making jokes about Aristophanes that you wouldn't understand.'

'That happens now!'

'This is serious.'

'I know . . .'

'Look, I've got something to show you. Hold out your arm, like this.' I held my right arm up, like a falconer calling his bird. Francis followed suit. 'See, over there!'

Two golden stars in the nebula were growing larger, heading towards us, gliding on their outstretched wings in the vacuum of space, fluttering into a stall to land on our arms, two magnificent phoenixes, reminiscent of peacocks but dull gold throughout apart from a broad golden circlet encrusted with diamonds around each neck.

Francis hadn't seen them before – I'd only just finished them.

'Oh, they're beautiful!'

'Do you like them? Do you see, I added diamonds to symbolize our sixtieth?'

'What's not to like? But what are they? I mean, I know they're phoenixes, but *why* are they?'

'They're the icons to our authentication protocols. You know, the code that tests who we are when we first log into the portal so no one else can use our avatars. I thought I'd make them a bit prettier.'

'Yes, yes, it's me you're talking to! Why don't you just admit you wanted to make your teenage fantasies of Rahylan and Avalon and the phoenixes come true?'

'Well, I thought it was a bit more romantic than making the icons look like a grey box.'

Francis softened. 'It is! But why now, when there's so much else to worry about?'

'At the moment, you have power of attorney for me. You can make life-or-death decisions on my behalf if I'm not able to. Legally, that ends when biological me dies. And the law is very unclear about who then can make life-or-death decisions on behalf of AI me. But it's *got* to be you, and no one but you. So I've set things up so that when biological me shuts down, AI me will simply pause and wait until you decide what you want to do.'

'I thought that was illegal if your AI really has become self-aware.'

'Firstly, I'm not covered by existing legislation because that only relates to stand-alone AI; there is no equivalent to the revised Turing test for cyborgs like me. Secondly, if AI me *is* self-aware, then he and biological me both agree that this is what we want for you.'

'What do you mean?'

'It's got to be *your* choice. Promise me you'll think it through very carefully . . .'

He gave me a kiss on the lips.

'I promise.'

Love Never Dies

Francis took a last look at the face he'd loved for over sixty-one years. He bent down and kissed its cheek, habitually, then pulled back, hesitated and kissed it on the lips for the first time for decades. He couldn't do any harm now. He pulled back again and took a last, last look. Peter's face in death looked exactly as it had for years. Maybe slightly paler.

And then he gave a slight nod, tears running down his cheeks; the strangers in the bedroom silently zipped up the black body bag and respectfully wheeled it out; he looked lovingly at Andrew and David; he turned away.

'Highcliff, I've decided to access Peter's AI. He said I should tell you.' Silence. Francis began to worry. 'Did he explain?' Silence. 'I *need* to solve The Bright-Eyes Riddle!'

'Of course, Francis. In which case, Peter left a final message for you.' Highcliff's tone was deeply respectful. 'He said: "Return to the obelisk where love never dies" . . .'

'Where is he, Tang?'

'I'm afraid the Archmage isn't online, Avalon.'

Francis found himself wondering if Tang's software was clever enough to know the truth and resort to euphemism, or if it simply didn't know.

'In which case, take me to the Flame of Analax.'

'I can, of course, but you know I can only take you to the edge of the clearing. The suns aren't up yet, so I'll give you a torch.'

The map tilted, the image darkened and suddenly prince and familiar were in near-darkness, a pre-dawn glow in part of the sky just sufficient to reveal they were by the edge of a forest at the base of a huge mound that might have been a grassy hill if it were not so perfectly symmetrical.

'I'll wait for you here.'

Francis looked briefly at Tangbone, clearly illuminated by the flaming torch he discovered his avatar was suddenly holding high in his right hand, then set off up the hill. It was a long climb, but as Avalon he could easily run up it. He didn't. Part of him was eager to get to the top; most of him dreaded what he would find. Or not find.

The huge mound had a flattened summit, so no one who climbed to the top could see what was there until their final steps. Francis knew this, and he slowed, then with new resolve speeded up for the final stretch. The sky to the west was glowing brighter. Francis was now above the treeline and he could see the horizon beginning to glow crimson. And then, looking ahead again, he realized he could see over the ridge.

This was where Avalon had first caught sight of Rahylan, meditating from sunset until sunrise on top of the obelisk, within the Flame. This was where they had irrevocably fallen in love. This, Francis assured himself, was where Peter would try to return. If it were possible. If anyone could. For him.

But the Flame looked empty.

Then again, the sky behind the Flame was still dark. Maybe there was a slight silhouette. Francis began running, then stopped, almost within touching distance, confused.

A hint of Rahylan sat in the Flame. Eyes shut. Transparent. An after-image. A memory. A reminder. A spectre.

Francis stood with his feet at the edge of the pit that surrounded the obelisk and made it apparently impossible for anyone to reach the Flame. He looked across the gulf, at eye level with the ghost he loved.

'Peter, are you there?'

The question hung unanswered.

'Are you there?'

He shouted it so loud that it triggered a long-unused, long-forgotten nuance of the AI and in the distance two woodpigeons startled by the noise took flight.

Then there was silence again.

Nothing moved. Not the Avalon avatar. Not the archived snapshot of the Rahylan avatar. A coral sky was now pushing away the dark. Unaware of the irrelevance of what it was doing, the software was preparing to create an impossibly beautiful sunrise. Francis was dimly aware of the dissonance, and in frustration threw down the now redundant torch, which fizzled and then went out in the computer-generated morning dew.

It was only as streaks of orange-red spread overhead and Francis remembered how, soon after they'd first met, Peter had explained that the ancient Greek poet Homer kept using the phrase 'rosy-fingered Dawn crept across the sky' that he finally broke down. On Earth, the sun was about to set; in Salania, one of two suns was about to rise; in both universes, tears rolled down his face.

'Goodbye, my darling . . .'

Any last words?

'I – Love – You – Peter.'

Eventually, he turned and slowly walked away.

Then he stormed back and raged at the illusion in front of him:

'You cheated death for twenty years – why couldn't you carry on?'

He screamed into a void of virtual reality:

'I don't want to be alone!'

This time he was unaware of whether a software algorithm generated a few more woodpigeons as the first rays of the first sun broke above the horizon. He was unaware of the manually coded Boolean logic Exclusive-Or function invoked somewhere in the cloud. He was unaware of the eyes opening. At least until he heard the familiar voice:

'I thought you'd never ask . . .'

Francis froze, simultaneously incredulous and hopeful.

'Are you still there?'

'Always! I'm yours forever.'

Was it a cruel trick? Was it a kind trick?

'Is it really *you*?'

Rahylan pulled a familiar face.

'I *think* so! But I'm not sure it matters . . .'

He suddenly stood up on the obelisk and without slowing his movement somersaulted forward across the pit to land in front of Avalon. He gave him a kiss. And they looked at each other.

'Wait a moment.' Francis broke the mood. 'Why didn't you answer immediately? I thought I'd lost you!'

'I'm sorry! You know how literal AI can be. It needed to be sure you were sure.'

For a moment Avalon's face relaxed. Then suspicious again:

'And what about the contrived dawn? The beautiful sky and the first sunrise *exactly* on cue?'

'The AI is learning how to be romantic. I suspect it would have delayed dawn forever until you *finally* made up your mind!'

'My God! It really *is* you! Whoever you are . . .'

They both laughed. And relaxed. Rahylan led the way across to the edge of the plateau.

'In the last few weeks, I've been exploring what we can do. Salania is changing. The whole portal is changing. Of course, more and more people are adding to it all the time. But there's something even more amazing going on. It's not just *my* AI that's evolving, it's *every* AI that you and I can access through the portal. Look, I want to show you something . . .'

He reached out and held Avalon's hands and without any fuss they were on another plateau, but this time high on a mountain, near the peak. Francis stopped, looked at the amazing view and then looked at his companion, then looked at his own arms then legs. He and Peter were both now dressed in what looked like tight-fitting VR suits – similar to the one an old man was wearing back in Torquay but a noticeable upgrade, a lot more spacey, the sort of thing people in the twenty-second century would wear. And Peter's haircut was different. Less Rahylan and more Peter at twenty.

'What the hell?'

'Welcome to our new world!' Peter waved his arms at the stunning vista.

'What's with the new look?'

'They're chameleon suits! They're turned off at the moment, but I've designed them to transform to match whatever setting we're in. This way, we'll *always* fit in wherever we go.'

'What do you mean "wherever we go"?'

'It's just like in the song, remember? This really is a world of "pure imagination"! Think about it, once we're through the portal, perception *is* reality! We're free.'

He was smiling broadly, his perfect white teeth catching the early-morning sun. Then he got serious, and his voice increasingly took on the tone of someone pleading for a life to be spared.

'We can live – I mean *really* live. We can go anywhere we want to go. We can be anyone – *anything* – we want to be.' He paused, then spoke more slowly: 'And we can do it for longer than we ever dared dream possible . . .'

Two phoenixes called overhead, one young and newly risen, one old and nearing its time. Francis looked at the incredible view as he and Peter walked to the very edge of the towering precipice and stood hand in hand, looking out on the almost unbearably beautiful alien landscape, watching twin suns rising out of a turquoise ocean into an impossibly perfect sunrise. Peter turned to him.

'You can join me. There's still time. I know you've always said it's not for you. But please, *please*, think about it now. You could get a brain–computer interface fitted, get the very best AI. Better than mine. And you're still healthy, you'll live long enough for *your* AI to meld with you. It'll be even more powerful than mine was. Don't you see, we can stay together.'

He reached out and held Francis by the shoulders.

'Always remember, I'm nothing alone. But whatever the multiverse throws at us, together we are *invincible*!'

He looked increasingly desperate as he got no response.

'I *know* you've never liked the idea of *you* taking the cyborg route, but I absolutely can't *bear* to lose you now. You're the only thing that gives me purpose.'

Somewhere in the cloud, the AI threw every ounce of logic into its final words, every emotion from sixty-one years of being in love, every lesson about what it meant to be human, every atom of its humanity:

'*Please*, my darling, don't *leave* me! What's the point of me cheating death if I have no reason to live?'

Francis looked long and deep into Peter 3.0's eyes.

'I thought you'd never ask . . .'